Tim Sandle

Produced by Microbiology Solutions

http://www.pharmamicroresources.com

DEDICATION

This book is dedicated to my wife, Jennifer, for all her kind support.

CONTENTS

About the author

Dr. Sandle is a chartered biologist and holds a first class honours degree in Applied Biology; a Masters degree in education; and has a doctorate from Keele University.

Dr. Sandle has over twenty-five years experience of microbiological research and biopharmaceutical processing. Dr. Sandle is an honorary consultant with the School of Pharmacy and Pharmaceutical Sciences, University of Manchester and is a tutor for the university's pharmaceutical microbiology M.Sc course. In addition, Dr. Sandle serves on several national and international committees relating to pharmaceutical microbiology and cleanroom contamination control (including the ISO cleanroom standards). He is currently chairman of the Pharmaceutical Microbiology Interest Group (Pharmig) LAL action group and serves on the National Blood Service advisory cleaning and disinfection committee.

Other books by the author include:

Saghee, M.R., Sandle, T. and Tidswell, E.C. (Eds.) (2011): Microbiology and Sterility Assurance in Pharmaceuticals and Medical Devices, New Delhi : Business Horizons

Sandle, T. (2012). The CDC Handbook: A Guide to Cleaning and Disinfecting Cleanrooms, Grosvenor House Publishing: Surrey, UK

Sandle, T. and Saghee, M.R. (2013). Cleanroom Management in Pharmaceuticals and Healthcare, Euromed Communications: Passfield, UK

Sandle, T. (2013). Sterility, Sterilisation and Sterility Assurance for Pharmaceuticals: Technology, Validation and Current Regulations, Woodhead Publishing Ltd.: Cambridge, UK

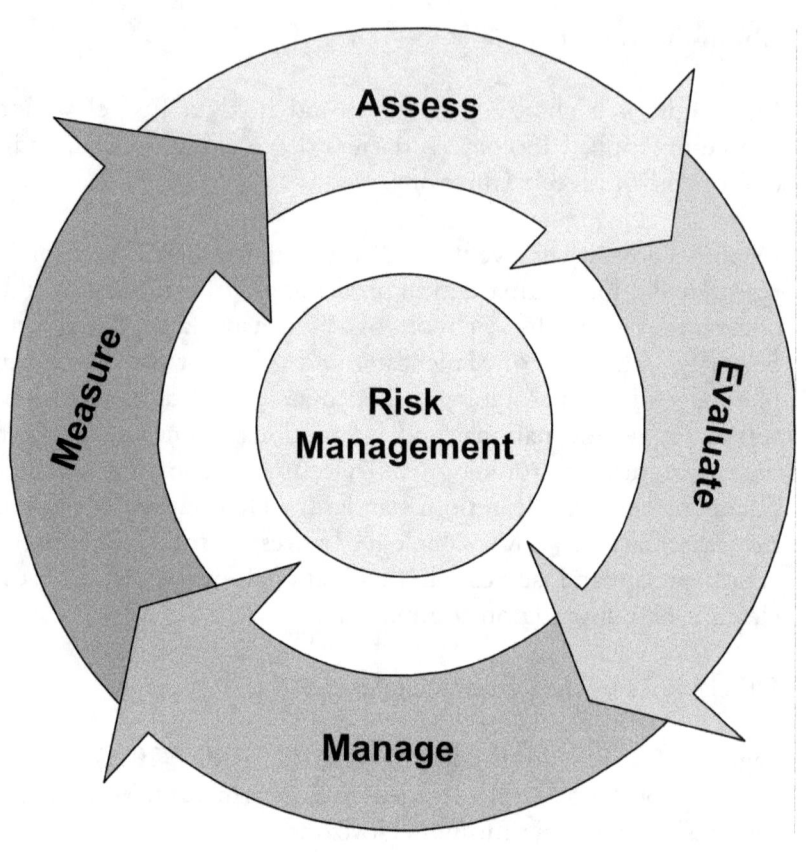

1 INTRODUCTION

This book presents an overview of risk management and risk assessment for those working in the pharmaceutical and healthcare sectors. An understanding of risk management and risk assessment is today becoming a prerequisite for those working in quality control and quality assurance, and for those active in pharmaceuticals and medical devices, Quality Risk Management it is a mandatory requirement. Although the book is aimed primarily at those involved with microbiology, sterility assurance or contamination control, the book also provides a sufficiently broad overview of the risk assessment tools and techniques to be of interest to the general reader.

Quality risk management is a systematic process for the assessment, control, communication and review of risks. It can be applied both proactively and retrospectively. The foremost concern of risk management is the protection of the product and the primary objective is the reduction of risks. This is achieved through the application of risk assessment tools. Amongst the most commonly used tools are:

- FMEA (Failure Mode Effect Analysis)
- FTA (Fault Tree Analysis)
- HAZOP (Hazard Operability Analysis)
- HACCP (Hazard Analysis and Critical Control Points)

These tools are explained and applied in the book.

The book is divided into six chapters. Chapter one provides an overview of Quality Risk Management and examines the various regulatory standards. Chapter two considers the primary risk assessment tools available, including the approaches outlined in ICH Q9 Quality Risk Management.

The book then proceeds to examine four case studies. Chapter three examines the application of risk assessment tools to plan and assess the microbiological and particulate monitoring of cleanrooms. Chapter four demonstrates how a HACCP (Hazard Analysis and Critical Control Points) approach can be used to assess the flow of materials and people (by using an example of an aseptic filling area). A second HACCP example forms the basis of Chapter five. Here HACCP is applied to a non-sterile production facility and the case study centers upon the receipt and testing of in-coming raw materials. Chapter six takes a different risk assessment tool and applies an FMEA (Failure Modes and Effects Analysis) to the design and operation of a sterility testing isolator.

2 INTRODUCING RISK ASSESSMENT AND RISK MANAGEMENT

Introduction

The use of risk assessment in the pharmaceutical industry is an expectation of regulatory authorities. Risk assessment forms a key component of any risk management program. The assessment of risk involves either the quantitative or qualitative determination of one or more risks. Risks are generally recognized as being related to a situation, event or scenario in which a recognized hazard may result in harm. Hazard, in this context, refers to any circumstance in the production, control and distribution of a (pharmaceutical) product which can cause an adverse health effect. This often refers to a biological, chemical or physical agent. Quantitative risk assessment requires a type of calculation. This is often based upon the magnitude or severity of the risk and the probability that the risk will occur (1).

Risk assessment involves identifying risk scenarios either prospectively or retrospectively (2). Prospective risk assessment involves determining what can go wrong in the system and all the associated consequences, and the extent or probability that it will go wrong. Whereas retrospective risk assessment looks at what has gone wrong and using risk assessment to assess the process, product or environmental risk and to aid in formulating the appropriate actions to prevent the incident from re-occurring (3).

Risk analysis is also highly beneficial in that it can also be used to identify and justify process improvements (4). Furthermore, the use of risk assessments can allow pharmaceutical manufacturers to explore weaknesses and to construct scientific and data based rationales. Risk assessment tools can also provide a means for the validation of processes (such as the approach referred to in the FDA Code of Federal Regulations, CFR 21, Part 820 (5).

The objective of this guide is to provide an overview of the basic principles of quality risk management (QRM) and to provide a framework for incorporating the risk management concepts in the quality systems thereby supporting the effective implementation of QRM. This is illustrated through several case studies which explore microbiological risks to products and processes.

Regulatory aspects

Quality Risk Management has been promoted by European medicines inspectors and by the FDA as an essential component of Twenty-First Century Good Manufacturing Practices (GMP). The most widely used GMP approach to quality systems is outlined in ICH Q10 "Pharmaceutical Quality System" (6). This current tendency towards a risk based approach has been spurred by three main guidelines:

1.   The 'Pharmaceutical cGMPs for the 21st century: A Risk Based Approach' initiative by FDA in 2002 (7).
2.   Annex 15 to the EU GMP Guide (8).
3.   The ICH Guideline titled Quality Risk management (ICH Q9) (9).

(The ICH-Q9 guideline concerning Quality Risk Management in the pharmaceutical field (active substances and medicinal products) was adopted by the European Union and PIC/S in Annex 20 of the EU and PIC/S GMP Guides.)

In addition, for medical devices, a specific guideline is in place. This is the European harmonized standard, EN ISO 14971:2000/ A1:2003, Medical Devices, Application of Risk Management of Medical Devices, supports the essential requirements of the European Directives related to risk.

Related approaches are:

• ISO/IEC Guide 73:2002 - Risk Management - Vocabulary - Guidelines for Use in Standards;
• ISO/IEC Guide 51:1999 - Safety Aspects - Guideline for their Inclusion in Standards; AS/NZS 4360:2004 - Risk Management;
• WHO Technical Report Series No. 908, 2003, Annex 7

Application of   Hazard Analysis and Critical Control Point (HACCP) Methodology to Pharmaceuticals;

- EN ISO 14971: Application of Risk Management to Medical Devices; Pharmaceutical Development (ICH Q8) and Annex (ICH Q8(R1));
- FDA Guidance for Industry PAT - A Framework for Innovative Pharmaceutical Development, Manufacturing and Quality Assurance;
- ASTM E2476 - 09 Standard Guide for Risk Assessment and Risk   Control as it Impacts the Design, Development, and Operation of PAT Processes for Pharmaceutical Manufacture;
- Pharmaceutical Quality Systems (ICH Q10);
- FDA Guidance for Industry Quality Systems Approach to Pharmaceutical cGMP Regulations.

The essentials of risk management

Risk assessment tools and techniques can be applied to every aspect of pharmaceutical processing. The expectation of QRM is to assess risks to the medicinal product and patient and then manage both to an acceptable level. An important part of this application involves an understanding of the process. Risk Management is fundamentally about understanding what is most important for the control of product quality and then focusing resources on managing and controlling these things to ensure that risks are reduced and contained. Before risks can be managed, or controlled, they need to be assessed (10).

Two important points to remember for any risk assessment are:

1.   There is no such thing as 'zero risk' and therefore a decision is required as to what is 'acceptable risk'.
2.   Risk Assessment is not an exact science - different people will have a different perspective on the same hazard.

Risk assessment should always:

- Be based on systematic identifications of possible risk factors,
- Take full account of current scientific knowledge,

- Be conducted by people with experience in the risk assessment process and the process being risk assessed,
- Use factual evidence supported by expert assessment to reach conclusions,
- Do not include any unjustified assumptions,
- Identify all reasonably expected risks-simply and clearly, along with a factual assessment and mitigation where required,
- Be documented to an appropriate level and controlled/approved,
- Ultimately be linked to the protection of the patient,
- Should contain objective risk mitigation plan.

Before embarking upon risk assessment it is important to establish and to define:

a)   Develop and agree upon the risk question. Clearly defining the risk question facilitates selection of the appropriate risk assessment tool, identifies relevant data, information and assumptions, and assists in the identification of resources, responsibilities and accountabilities.

b)   Select the most appropriate risk assessment tool. The selected method will be used to organize collected data, understand what steps can be taken to reduce or control risk and to help make appropriate decisions. Risk assessment methods are outlined in this guide, together with practical examples of their application.

Formal risk approaches normally share four basic concepts, which are listed below:

1)   Risk assessment,
2)   Risk control,
3)   Risk review,
4)   Risk communication.

These four steps can be interpreted as:

Risk assessment

Risk assessment is the assessment of effects of the incident or scenario (such as a microbiological contamination event) (11). It involves the use of risk analysis tools and the evaluation of risk. The object of using these tools is to identify hazards and the root causes of such hazards. Hazards include, but are not limited to:

• Materials and ingredients;
• Physical characteristics and composition of the product;
• Processing procedures;
• Microbial limits;
• Premises;
• Equipment;
• Packaging;
• Sanitation and hygiene;
• Personnel, such as human error; and
• Risk of explosions.

In undertaking risk analysis the following key questions should be addressed:

• What is the nature of possible hazards?
• What is the probability of their occurrence and how easy is it to detect them?
• What are the consequences (the severity)?

Risk control

Risk control is centered on risk reduction or risk mitigation. This process consists of corrective actions taken to resolve the incident and preventative actions proposed to avoid a recurrence of the incident in the future. At some point, risk control will also involve a consideration of risk acceptance for the measures put in place will ultimately need to be accepted or rejected.

Risk review

The risk review is the follow up of action items and the final summary and evaluation of the incident. This should be approved by senior management. In drawing the risk conclusions together for a risk assessment, the mitigation controls should minimize the likelihood of risk to patient safety to an acceptable level of assurance. Here, risk mitigation plans must be identified and implemented where any risk to patient safety is posed.

Following the completion of a risk assessment, appropriate systems should be in place to ensure that the output of the risk assessment process is periodically monitored and reviewed, as appropriate, to assess new information that may impact on the original risk management decision.

<u>Risk communication</u>

Risk communication discusses the important steps involved in reporting and discussing the incident. It should also ensure that the appropriate parties are included in resolving the incident. Communication of the risk process must include all key stakeholders within the organization.

<u>Conducting risk assessments: the process</u>

Before commencing a risk assessment it is important to define the magnitude and the scope of the assessment (remaining focused on what is to be achieved); to select the appropriate team (often an interdisciplinary team is best, and, in the context of this guide the microbiologist is essential); selecting and reviewing the appropriate risk management tool; deciding upon any numerical scale to be used (if any is applicable) and prioritizing the different problems to be addressed. Most approaches begin by constructing a process map.

The various analytical tools used for conducting risk assessments are similar, in that they involve:

- Constructing diagrams of work flows (these are pictorial representations of a process designed to break the process down into its constituent steps, this allows complex processes to be simplified);

- Identify hazards (which can be intrinsic, that is specific or inherent to the process or equipment, or extrinsic, that is factors which are external to the process or equipment but which might impact upon it).   In doing so, this helps to:
- Pin-point the areas of greatest risk;
- Allow an examination of the potential sources of contamination;
- Decide on the most appropriate sample methods;
- Help to establish alert and action levels for on-going monitoring and to detect future risks;
- They should be flexible enough to take into account changes to the work process and any seasonal activities.

In doing so a standard series questions should be asked. For example, when considering equipment breakdown:

- What is the function of the equipment? What are its performance requirements?
- How can it fail to fulfill these functions?
- What can cause each failure?
- What happens when each failure occurs?
- How much does each failure matter? What are its consequences?
- What can be done to predict or prevent each failure?
- What should be done if a suitable proactive task cannot be found?
- If a risk cannot be eliminated then how can it be reduced?
- If the risk cannot be reduced then how can it be monitored?

Thus the general approach is to recognize a risk, rate the level of the risk and then set out a plan to minimize, control and monitor the risk. Subsequent monitoring of the risk will help to determine, any follow up action.

There are many advantages of risk assessment equally here are also some 'disadvantages' (or at least pitfalls to be aware of). The advantages of using risk management methodologies include:

1.   The fact that decision making is improved and streamlined.
2.   The philosophy of risk management is applicable across the organization.   Risk management can be applied to all

business processes, processes related to product manufacturing and distribution, at different levels of a company and for regulatory authorities.

3. Activities can be prioritized; organizations can concentrate effort on what will be of greatest importance to the final product and therefore yield the greatest benefit to the patient population. This also helps with effective resource allocation.

4. The scientific and data driven nature of risk assessment methodologies significantly reduces subjectivity.

5. Risk assessment allows the organization to enhance its credibility with regulatory agencies.

6. Risk assessment provides one means of building quality into an organization.

7. Risk awareness becomes a behavioral trait. The introduction of a risk management concept for sustainability of operations requires that all employees involved are aware of the potential hazards involved and how to control risks.

8. Risk management helps to reduce overall cost: supports more qualified decision making in the planning stage.

9. Risk management promotes quality, through increased efficiency and knowledge transfer, with strong potential to reduce catch-up work done to mediate the effects of poor quality (i.e.: non-conformances, deviations/investigations, CAPA, rework, scrap, complaints, etc.).

10. Risk management provides a mechanism for risk communication (formalized vehicle/process) and exposure to management.

11. Risk management is a diagnostic tool in that it provides a framework to better understand processes, what is critical and why.

12. Risk management helps to provide rationale for not spending time on low risk activities, process events, or systems, rather focusing resources and time on the things that are really important.

There are, however, some disadvantages with risk management approaches. Many risk assessment tools are subjective and rely, to an extent, upon supposition. Often with a contamination event there can be multiple risks and some of the risk methods are weak when dealing with multiple risk situations. Additionally some techniques are not effective when used to filter out multiple risk factors (an example here is the Hazard Analysis and Critical Control Points method, which tends to be applied in a relatively linear way).

Furthermore, with some quantitative risk assessments there is a danger that qualitative differences between different risks are ignored. To compound this some assessment tools may drop out important non-quantifiable or inaccessible information. A further weaknesses arises with risk assessment tools still being relatively new within the pharmaceutical industry, which means there is only a small number of published case studies on which to draw upon.

In addition to the disadvantages, errors can occur when risk management is not executed properly. Common errors include:

- Failing to understand the difference between risk assessment (the individual documents) and risk management (the holistic process),
- When the tool used is too complex,
- Lacking of planning,
- Unclear definitions,
- Unclear qualitative grading,
- Not undertaking the exercise in a timely manner,
- When the outcome is already decided,
- When all of the required process knowledge is not available,
- When the team used for the risk assessment is not multifunctional/multidisciplinary,
- When combinations of tools not considered,
- When the outcomes of the risk assessment are not liked and changed to suit what senior management might expect,
- Where there has been insufficient review and communication.

Nevertheless, with these weaknesses understood the advantages of risk assessment outweigh the disadvantages, as the case studies in this guide illustrate.

Summary

In drawing the key elements together, as well as the identified risk methodologies, risk management should be approached in a way which involves subject matter experts. Once the risk assessment exercise has been completed, the outcome must be communicated between the decision maker and to all stakeholders. At later stage, the risk assessment should be reviewed to see that it is still appropriate.

Risk assessment allows processes and equipment to be designed in ways which are more efficient and safe for users and patients, and risk assessment allows problems which occur during processing to be addressed so that the patient can be safeguard and preventative actions formulated to prevent a re-occurrence.

References

1. Vesper, J.L. (2006). Risk Assessment and Risk Management in the Pharmaceutical Industry: Clear and Simple, Davis Healthcare International Publishing, LLC, USA.

2. ISO/IEC Guide 73:2002: "Risk Management - Vocabulary", Guidelines for Use in Standards", International Standards Organisation, Geneva, Switzerland.

3. Kemppainen J.K. (2000): The critical incident technique and nursing care quality research. J Adv Nurs; 32(5):1264-71.

4. Sidor, L. and Lewus, P. (2007): 'Using risk analysis in Process Validation', BioParm International, pp50-57

5. Code of Federal Regulations, Title 21, Food and Drugs, Part 820, Quality Management Regulations, 820.75

6. International Conference on Harmonisation of Technical Requirements for Registration of Pharmaceuticals for Human Use, ICH Harmonised Tripartite Guideline, Q10: Pharmaceutical Quality System, 2008

7. US Food and Drug Administration Pharmaceutical GMPs for the 21st Century – A Risk Based Approach. Final Report, Rockville, MD. September 2004

8. Euradlex. The Rules Governing Medicinal Products in the European Community, Annex 1, published by the European Commission

9. International Conference on Harmonisation of Technical Requirements for Registration of Pharmaceuticals for Human Use, ICH Harmonised Tripartite Guideline, Q9: Quality Risk Management, Version 4, 2005

10. Ingram, D (2009). "Technical Problem Solving", Journal of Validation Technology, Winter 2009, pp. 64 – 70.

11. Whyte, W. and Eaton, T. (2004): 'Microbiological contamination models for use in risk assessment during pharmaceutical production', European Journal of Parenteral and Pharmaceutical Sciences, Vol. 9, No.1, pp11-15

3 APPROACHING RISK ASSESSMENT: TOOLS AND METHODS

Introduction

The current environment in the pharmaceutical industry and within the healthcare sector is influenced by the challenge of an appropriate balance between increased compliance requirements with GMP/GDP, regulatory guidance and legal enforcements, versus the use of risk assessment. Quality Risk Management allows process and products, and in some cases personnel, to be better protected and it can be used to help to meet regulatory expectations and to meet cost demands or to seek process efficiencies (Phoenix and Andrews, 2003).

Risk management is not a one-off activity, and whilst it is useful as a tool for responding to problems, it is best seen as a proactive tool for avoiding risks from occurring. Risk management is integral to Quality by Design and Product Lifecycle philosophies.

The aim of risk management is to systematically assess, control, and review manufacturing processes in terms of priorities, and subsequently develop appropriate measures to control risks. For this there are a variety of different ways with which to undertake risk assessment. The use of more systematic tools represents the shift away from tick box approaches to more scientific assessments of risk, such as those encouraged in the US FDA document "Pharmaceutical cGMPs for the 21st Century: A Risk-Based Approach" and in the ICH Q9 Quality Risk Management guideline. Other similar approaches are: ISO/IEC Guide 73:2002 - Risk Management - Vocabulary - Guidelines for use in Standards; ISO/IEC Guide 51:1999 - Safety Aspects - Guideline for their inclusion in standards; AS/NZS 4360:2004 - Risk Management; WHO Technical Report Series No 908, 2003, Annex 7 Application of Hazard Analysis and Critical Control Point (HACCP) methodology to pharmaceuticals; EN ISO 14971: Application of risk management to medical devices;

Pharmaceutical Development (ICH Q8) and Annex (ICH Q8(R1) and FDA Guidance for Industry PAT - A Framework for Innovative Pharmaceutical Development, Manufacturing and Quality Assurance; Pharmaceutical Quality Systems (ICH Q10) and FDA Guidance for Industry Quality Systems Approach to Pharmaceutical cGMP Regulations

The approaches to risk assessment normally centre on identifying a risk, and then by assessing the extent to which the risk would be a problem (the severity of the risk) and secondly by calculating how likely it is that the risk would occur (the probability). Following this, attempts are made to mitigate the risk (ideally eliminate the source or take measures to reduce the likelihood of it happening). Where the possibility of the risk occurring can be reduced no further, then steps should be taken to monitor the risk (detection), preferably as an early warning, so that action can be taken.

The above constitutes a proactive approach to risk assessment. Risk assessment can also be reactive in terms of formulating a response should an event which appears to represent a risk occur. In terms of pharmaceutical processing this could be the failure of an HVAC system during processing. Here a variety of scenarios could be adopted, from additional testing through to batch rejection.

The final parts of the approach to risk assessment are documenting the risk assessment activity and in communicating the findings. The latter is important for risk assessments are of little value should they reside on shelves gathering dust. They need to be used and regularly reviewed and updated as necessary.

This chapter focuses on risk assessment and explores some of the different tools and techniques available for conducting risk assessments. Whilst the examples relate to the pharmaceutical and healthcare sectors none of the techniques originate from these sectors. Instead they were devised and developed long before in the engineering industry (much centred on the post-World War II Total Quality movement associated with Deming) (Deming, 1986); and from food and beverages (as initiatives to prevent costly rejections through microbial contamination). Furthermore, much of the terminology commonplace with risk assessment (such as "hazard") is derived from industrial health and safety legislation.

Approaching risk assessment

There are two groups of approaches to the risk analysis process. These are qualitative and quantitative methods. Of these, the qualitative a approach is the more simple and involves the identification of risks to which a process or a procedure is exposed. For the process the optimal approach is to gather the relevant expert together, come up with a list of risks and then qualify which risks are worth acting upon. This process is more intuitive and is enhanced by answering three basic questions (Jackson, 2001):

• What could happen?
• How likely is it to occur?
• What is the impact?

The downside with the qualitative approach is that it is difficult to rank risks relative to each other or to place the risks in context. It is also difficult to assess the effectiveness of an action or to measure if a risk previously assessed at a 'high level', for example, has decreased to a lower level.

Quantitative approaches normally involve the use of numbers so that one risk can be ranked or measured against another and compared to a pre-determined scale. Such an approach allows a decision to be taken about the cost-benefits of investing in a risk reduction strategy and to allow the costs of monitoring risks relative to the likelihood of the risk occurring to be taken.

The methods of risk management outlined in ICH Q9, Annex 1 are:

1. Basic risk management facilitation methods
2. Failure Mode Effects Analysis (FMEA)
3. Failure Mode Effects and Criticality Analysis (FMECA)
4. Fault Tree Analysis (FTA)
5. Hazard Analysis and Critical Control Points (HACCP)

6. Hazard Operability Analysis (HAZOP)
7. Preliminary Hazard Analysis (PHA)
8. Risk ranking and filtering
9. Supporting statistic tools

Both qualitative and quantitative approaches to risk assessment involve the identification of hazards. A hazard is a situation that poses a level of threat (in the pharmaceutical and healthcare context. A hazard is any biological, chemical, mechanical, or physical agent). Generally, most hazards are dormant, with only a theoretical risk of causing 'harm' to people, product, process or to the environment. Hazards and risks mean different things. A hazard is an ever present property. The definition of risk is a combination of consequences (severity) and their probability When hazard and vulnerability interact together this creates "risk" and when something occurs this is often described as an "incident" or "event" (with pharmaceutical processing this could be the 'deviation', out of specification result of microbiological data deviation). This is relationship between hazards and risk is commonly summarized in the formula:

Risk [R] =

Likelihood of Occurrence [S] x Seriousness if incident occurred [F]

Or:

R   = S X F

Sometimes likelihood is referred to as the frequency and seriousness as the severity.

In pharmaceuticals and healthcare, the main risks fall into the following groups:

• Risks from facilities e.g. HVAC systems
• Risk from personnel, particularly contamination
• System risks such as new regulatory compliance
• Process risks, such as contamination affecting the process

• Product risks, based on the failure to meet specifications.

Once a hazard has been identified it needs to be assessed. This involves weighing up how significant the risk is. The assessment should centre on eliminating the hazard, reducing the potential for harm (risk control) and / or monitoring it.

A second commonality of risk assessment approaches is root cause analysis. Root cause analysis is similar to critical incident technique, and is s a formalized investigation and problem-solving approach focused on identifying and understanding the underlying causes of an event as well as potential events that were intercepted (Kemppainen, 2000).

The monitoring aspect is a fundamental part of risk management. Reducing the potential for harm way involve strategies like the re-design of equipment, a change to personnel clothing or changing a work flow (risk management should be seen as part of the Quality by Design approach). The introduction of PAT (Process Analytical Technologies) in the 2000s has introduced technologies that enable more process parameters during manufacturing to be monitored whilst the process is underway (Tan et al, 2005).

The most important aspects of risk management, as defined by ICH Q9, are:

- Identification,
- Analysis,
- Evaluation,
- Control by reduction measures and risk acceptance,
- Communication and
- Monitoring/Review

The most effective risk assessment tools allow these key aspects to be captured, processed and examined.

Simple risk assessment tools

There are a host of risk assessment tools available. Many overlap and some are more useful for different situations than others. This chapter focuses on some of the most common (and easy to use) approaches.

Although this section is entitled "simple risk assessment tools" this does not mean inferior but in recognition that some situations require more sophisticated risk tools than others. In some situations a simple check-list, for example, will suffice.

Check-lists

A check sheet is a structured, prepared form for collecting and analyzing data. This is a generic tool that can be adapted for a wide variety of purposes. A tally chart for observing events is an example.

Fishbone Diagrams

Cause and Effect Diagrams, tools for quality improvement, are also known as Fishbone Diagrams because a completed diagram can look like the skeleton of a fish; and sometimes as Ishikawa Diagrams (named after Professor Kaoru Ishikawa (1915-1989), a pioneer of quality management, who devised them in the 1960s).

The first cause and effect diagrams were applied to industrial situations and were based on five Ms: Manpower, Methods, Materials, Machinery, and Milieu [the physical or social setting in which something occurs or develops, sometimes referred to in text books as 'Mother Nature'] and were used to observe and visualise cause/effect relationships. The 'Ms' can be used or replaced by other factors applicable to the area being investigated. An example is shown in the diagram below:

Fishbone Diagram

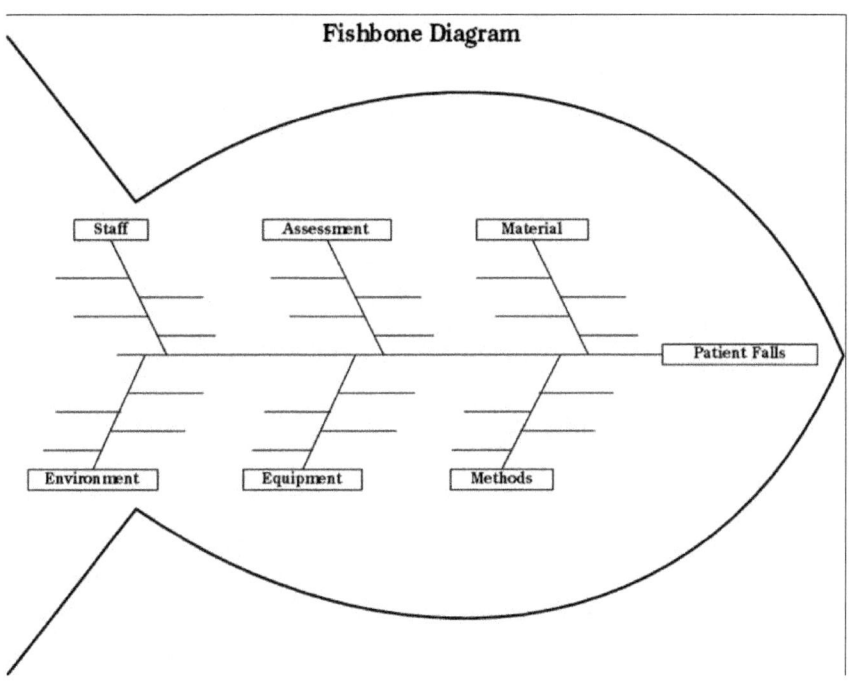

The fishbone diagram begins with a horizontal arrow pointing to the right, representing the backbone of the fish. The problem is formulated as concisely as possible at the head. The cause groups (which can be then five Ms or related 'causes') then extend upwards and downwards along this main spine. A variety of creativity techniques are used to identify the cause of the individual problems. Thus the fishbone analysis provides a structured way to help the users think through all possible causes of a problem and to carry out a thorough analysis of a situation. This is undertaken through the following steps:

1. Identify the problem:

For step 1 the specific problem should be defined in detail. At this stage it is useful to identify who is involved, what the problem is, and when and where it occurs. Write the problem in a box on the left hand side of a large sheet of paper. Draw a line across the paper horizontally from the box. This arrangement, looking like the head and spine of a fish, gives the space to develop ideas.

2. Work out the major factors involved:

For step 2, identify the factors that may contribute to the problem. Draw lines off the spine for each factor, and label it. These may be people involved with the problem, systems, equipment, materials, contamination sources and so on. It is good practice to draw out as many possible factors as possible (these can always be eliminated later on). Many of the best thoughts and ideas arise from group discussion ('brainstorming').

3. Using the 'Fishbone' analogy, the factors found should be drawn like the bones of the fish.

4. Identify possible causes:

For each of the factors considered in step 2, all of the possible causes of the problem should be considered. These are shown as smaller lines coming off the 'bones' of the fish. Where a cause is large or complex, then it may be best to break it down into sub-causes. Show these as lines coming off each cause line.

5. Analyze the diagram:

By stage 5 there should be a diagram showing all the possible causes of the problem. Depending on the complexity and importance of the problem, the most likely causes can be investigated further. This may involve setting up investigations, carrying out audits, discussing specific issues with staff and undertaking additional testing. Many of these should be designed to test whether your assessments are correct. In some cases it is a good idea to have a pre-defined protocol drawn up to avoid reacting to the findings in a biased way. For example, if additional testing is carried out and the results are positive or negative, the responses to these should ideally have been considered before the testing is undertaken.

Some useful areas to examine when using fishbone diagrams for problem solving are:

A possible start can be to answer the question "Which factors can influence the quality of the product?" as bases for the main branches, which can be formed as follows:

- Personnel, e.g. employees approach to contamination control, basic   hygiene, areas where manual operations are required or open   processing occurs.
- Machinery, e.g. equipment in direct contact with the product, likely   equipment breakdowns, malfunctions (such as equipment generating   particles)
- Materials, e.g. starting material, packaging material, raw materials
- Methods, e.g. manufacturing process, test methods and results
- Environment, e.g. failure to meet operational parameters such as HVAC failure
- Management, e.g. inadequate procedures

The main advantage of the fishbone diagram is that all influencing variables are displayed, systematically, on one page. The main disadvantage is when studying more complex systems and thus fishbone diagrams are more effective for analyzing a single step rather than a complex processes.

Contradiction matrix

A contradiction matrix can be used either as a standalone problem solving tool or in conjunction with the fishbone diagram (Ingram, 2009). The contradiction matrix allows facts and causes to be examined to see if they are relevant.

For example:

Potential Causes	Established Facts			
	Fact 1	Fact 2	Fact 3	Fact 4
Cause 1	O	O	X	A
Cause 2	N/A	X	O	O
Cause 3	X	O	O	X
Cause 4	X	N/A	O	N/A

Where:

    O   =   Cause is supported;
    X   =   Cause is contradicted;
    A   =   Cause is supported with assumptions;
    N/A =   No relationship between cause and effect.

Is / Is Not method

Similar to the contradiction matrix is the "Is / Is Not" method. This provides a formal means for capturing the process of elimination and is useful to show regulators or audits so that the audit question of "did you think of x?" can be simply answered "yes we did but we did not think it was important".

The process of elimination involves considering experience and knowledge, weighing up events and timelines, considering circumstances and alternatives. This documentation normally includes:

• A description of the circumstances and outline conditions as well as a question (formulated as concisely as possible)
• A brief list of possible alternatives or solutions
• An assessment of the possible risk ("What could happen, if ..." or "What could happen, if.. not...")

A good way to capture this is by using an IS/IS NOT table for looking at each cause or fact. Things which fall in the 'is not' column need not be examined further whereas things that fall into the 'is' column are suspect and should be explore further.

An example template is:

Potential cause or fact	IS	IS NOT
What		
When		
Where		
Extent (occurrence)		
Severity		

An IS/IS NOT table is designed to help establish facts. It can be expanded or contracted as required.

Applying the contradiction matrix or is / is not method

In order to gather information for the contradiction matrix or is / is not table, the following methods are useful

• Trend analysis, including SPC and control charts (of which Shewhart, cumulative sum and the exponentially weighted moving average are useful). With all trend charts, data distribution and data transformation should be considered.

• Process flowcharts, e.g. how a product is made; how people or material move. These are useful for creating process maps.

Flowcharts

A flowchart is a diagram that represents a process (the idea is to visualize what is going on). The chart displays the steps as boxes of various kinds, and their order by connecting these with arrows. This diagrammatic representation can give a step-by-step solution to a given problem. Process operations are represented in these boxes, and arrows connecting them represent flow of control.

An example flowchart is:

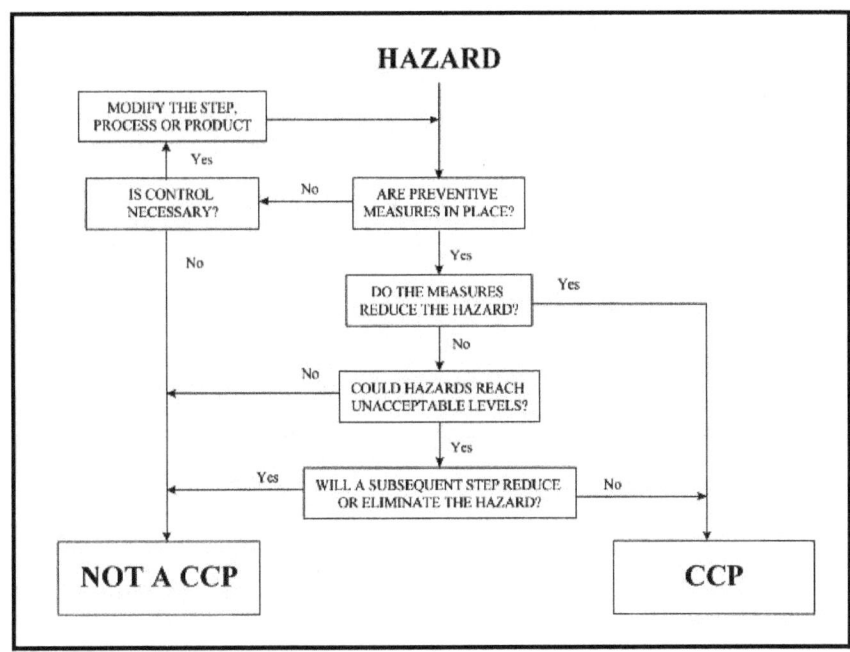

The process flow method (above) shows how actions are interrelated and is able to also integrate interfaces into the flow. It provides a clear and simple visual representation of involved steps. Therefore, this method facilitates understanding, explaining and systematically analyzing complex processes and associated risks.

For the flow chart, standard 'flow chart symbols' are used. Although a range are available it is best that these are limited otherwise there is a risk of obfuscating the overall picture. For example, a processing step, usually called activity, and denoted as a rectangular box and a decision is usually denoted as a diamond.

Examples of flowchart symbols are shown below:

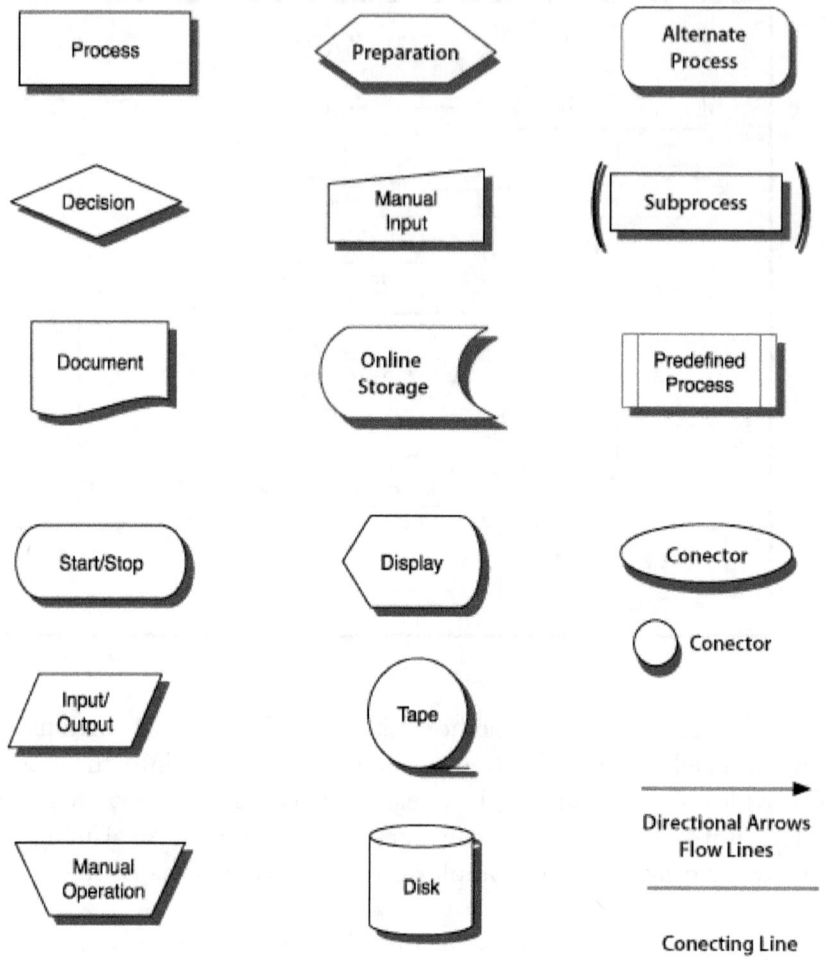

Process

Preparation

Alternate Process

Decision

Manual Input

Subprocess

Document

Online Storage

Predefined Process

Start/Stop

Display

Conector

Conector

Input/ Output

Tape

Directional Arrows Flow Lines

Manual Operation

Disk

Conecting Line

Advanced risk assessment tools

More advanced risk assessment tools are useful for looking at more complex problems. Although such tools can prove to be very useful they take time construct and should not automatically be the default option for a simple flowchart or fishbone diagram can sometimes be just as effective.

The common risk assessment approaches each has advantages and disadvantages, which are summarised in the table below:

Focus	Methods	FTA	FMEA	HACCP	Statistical methods
Risk assessment	Risk identification	+	o	o	+
	Risk analysis	o	+	o	+
	Risk evaluation	-	+	o	+
Controlling risks	Risk reduction	-	+	+	-
	Risk acceptance	-	+	+	-
	Risk review	A single method may or may not be ideal, sometimes different approaches need to be combined.			
	Risk communication				
(+ = very suitable, o = limited suitability, - = not suitable)					

Advanced risk assessment techniques involve the assembly of problem solving teams for the risk assessment process. The objectives of the team should be to:

Define the scope

- Describe the interfaces affected by the defined scope e.g. on system, process, environment, facility or products
- Have a clear description of problems or questions, which the risk management should answer
- Selection the risk assessment method(s)
- Define the acceptance criteria
- Define a team including roles and responsibilities
- Collect all relevant data e.g. internal knowledge and regulatory requirements.

Fault tree analysis

Fault Tree Analysis (FTA) is an established technique (see for example CEI IEC 61025 and DIN 25424) and it is widely used in the engineering industry, particularly for safety management. Essentially a fault tree is a logical diagram which shows the relation between system failure, that is a specific undesirable event in the system, and failures of the components of the system. Thus FTA provides a graphical depiction of all causal chains of failure of a system or sub-system.

FTA is often termed a 'top down' approach and works well if the correct failure has been identified for FTA assumes failures follow from logical combinations of causes that occur with known probability.

The five steps for undertaking FTA are:

1. Define the problem

Here the event should be outlined and agreed. At this stage brainstorming is useful, based around the question "why has this event happened?"

2. Define fault criteria

This step involves defining what has caused the event to occur, such as a deviation from a specification.

3. Define the specification

Here the specification and the reliability of the measure should be reviewed to ensure that they were satisfactory.

4. Evaluate causes

At this step the different causes should be considered. These should be categorised into similar types, such as technical faults, human error, equipment failure, loss of environmental control and so on.

5: Draw the fault tree

The fault tree should start with the event, followed by any other events and then the possible causes (casual relationships of the failures leading to the event). The fault tree can be 'top down' or of a 'left-right' design.

An example of a fault tree is:

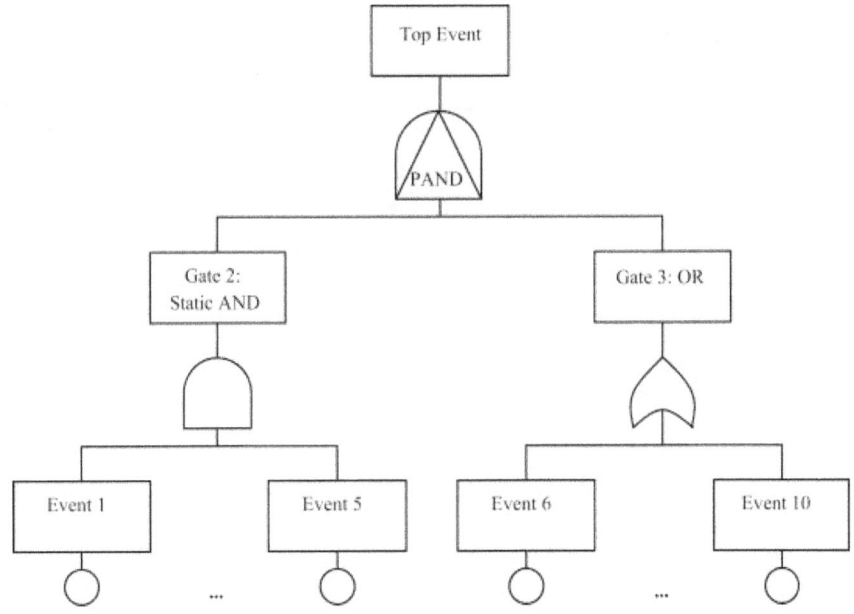

At each level in the above tree, combinations of fault modes are described with logical operators (AND, OR, etc.). Each established cause is investigated until no more answers can be found. This method results in a fault tree with a varying number of branches and sub-branches depending on its complexity.

Where FTA can go wrong is if the wrong cause is selected and the sub-branches fail to detect the actual issue. Furthermore, FTA is a better retrospective analytical tool rather than as preventative measure.

Hazard and Operability studies (HAZOP)

The HAZOP (Hazard and Operability studies) technique was also developed in the engineering sector . HAZOP involves a formal systematic critical examination of process in order to assess the hazard potential that arise from deviation in design specifications and the consequential effects on the facilities or system as a whole. HAZOP is based on a theory that assumes risk events are caused by deviations from design or operating intentions.

This technique is usually performed using a set of guidewords, such as:

- NO/NOT
- MORE OR/LESS OF
- AS WELL AS,
- PART OF
- REVERSE,
- AND OTHER THAN

From these guidewords, scenarios that may result in a hazard or an operational problem is identified. For this, the consequences of the hazard and measures to reduce the frequency with which the hazard will occur can be discussed.

Many HAZOP approaches utilize process flow diagrams (similar to the type discussed above). From this, a HAZOP table is constructed, deploying some or variants of the guidewords above. For example (BSI, 2002):

Guideword	More	Less	None
Flow	high flow	low flow	no flow
Pressure	high pressure	low pressure	vacuum
Temperature	high temperature	low temperature	
Time	too long / too late	too short / too soon	sequence step skipped
Utility failure (instrument air, power)			failure

Reverse	As well as	Part of	Other than
reverse flow	deviating concentration	contamination	deviating material
	ΔP (delta-p)		explosion
backwards	missing actions	extra actions	wrong time

The advantage of HAZOP is that it can be good add detecting and handling hazards which are difficult to quantify, such as hazards rooted in human performance and behaviors or hazards that are difficult to detect.

The disadvantage is that there is no means to assess hazards involving interactions between different parts of a system or process and there is no option for risk ranking or prioritization.

Failure Mode and Effects Analysis (FMEA/FMECA)

Failure Mode and Effects Analysis (FMEA) is a further technique which originated from the engineering sector . FMEA is a process whereby each potential failure mode in a system is analysed to determine its effect on the system and to classify it according to its severity. "Failure modes" means the ways, or modes, in which something might fail. Failures are any errors or defects, especially ones that affect the customer, and can be potential or actual (Tague, 2004).

By looking at each failure mode FMEA methodically breaks down the analysis of complex processes into manageable steps.

FMEA is a preventative method that provides a quantitative evaluation (through numerical scoring) of potential failures (faults) and their potential causes in terms of:

- Severity of the failure (S)
- Probability of occurrence (P)
- Probability of detection (D)

These three factors are normally multiplied together to give a risk priority number (RPN). The bigger the number then the bigger the potential risk and, it follows, the greater the priority and resources required to address the risk.

Not all FMEA models use detection. Detection is, however, more often a valuable tool when considering probability.

Each potential failure mode is separately assessed.

Before undertaking an FMEA the scoring system must be decided in advance. This is normally a scale of 1 - 5 or 1 -10 for each of the three causes above. Based on this, with a scale of 1 to 5, a high severity event (such as a product contaminated with endotoxin) would be given a score of 5 where a low severity event (such as minor reduction in temperature for a very short period of time) would be given a score of 1. With probability, if something is very certain to happen (such as processing without a HEPA filter running) then a score of 5 would be given, whereas if something was very unlikely to happen (such as forgetting to filter a process step when this is stated on a batch record step) then a score of 1 would be given. With detection, if there is a good detection system in place a score of 1 is given (as this weights the final score downwards) whereas a non-existent detection system would be given a score of 5 (to weight the final risk priority number upwards).

Some examples of rating scales are:

Severity rating scale

Rating	Criteria: A failure could...
10	Harm a customer or employee.
9	
8	Render the product unfit for use.
7	
6	Result in partial malfunction
5	
4	Cause minor performance loss.
3	
2	Be unnoticed; minor effect on performance.
1	

Occurrence rating scale

Rating	Time period / probability
10	More than once a week >30%
9	
8	Once per week ≤5%
7	
6	Once every 3 months ≤0.03%
5	
4	Once per year ≤6 per 100,000
3	
2	Once every 3-6 years ≤3 per 10 million
1	

Detection rating scale

Rating	Definition
10	Defect caused by failure not detectable
9	
8	Units are systematically sampled and inspected
7	
6	Manual inspection with mistake proofing modifications
5	
4	Immediate reaction to out of control conditions
3	
2	All units are automatically inspected
1	

The above tables include some examples for each rating number. Other factors can be filled in as appropriate.

With the calculated risk priority number there is often some kind of cut-off value whereby each FMEA above the value must be addressed as a potential major risk whereas FMEA outcomes below are a lower risk and do not require immediate action. For example, using the 1-5 scale, the selected cut-off value might be 27 (based on 3 (medium severity) x 3 (medium probability) x 3 (medium detection)). Instead of a numerical system, some elect to use decision trees with categories of low-medium-high (McNally et al, 1997).

The primary steps are:

1. Collect Basic data
2. Describe process conditions
3. Hazard identification e.g. identification of possible failures, consequences and cause of failure
4. Hazard assessment (Risk Analysis)
5. Evaluation of the failure and determination of the risk priority number (RPN)
6. Definition of reductions measures
7. Awareness of the residual risks
8. Summary of the results
9. Documentation of the performed process
10. Follow up and the implementation of measures

When the FMEA is extended by a criticality analysis, the technique is then called failure mode and effects criticality analysis (FMECA). FMECA is an extension of FEMA which allows each failure mode to be identified and then evaluated for criticality. This criticality is then translated into a risk, and if this level of risk is not acceptable, corrective action should be taken.

FMEA can be very effective and it works well as a preventative risk assessment tool. One main advantage is that is allows risks to be ranked and action prioritized. However, the identification of possible failures is a time consuming task.

An example of an FMEA is shown below. This relates to a pharmaceutical manufacturing operation. In the example priorities have been identified from the analysis by looking at the potential failure modes with the highest Risk Priority Number (RPN). This figure is obtained by multiplying Severity (SEV), Occurrence (OCC) and Detection (DEC).

Process Step/Input	Potential Failure Mode	Equipment	Potential Failure Effects	SEV	Potential Causes
	Incorrect Flow path	Unicorn Control System	Incorrect operation	8	Error in process method. Method compromised/ corrupted.
	Software Failure	Unicorn Control System	Incorrect operation	6	Software compromised/ corrupted.

Potential Causes	OCC	Current Controls	DET	RPN	Actions Recommended
Error in process method. Method compromised/ corrupted.	1	Changes to method restricted to administrator access level and audit trail of operations built into software. Individual logging onto system with defined operations for each level/operator. System locks when no operation performed within given time period. Process methods checked pre-use.	1	8	Three buffer runs to confirm process operation against method. First three product runs monitored against current parameters.
Software compromised/ corrupted.	1	Routine process monitoring. Analysis of computer printout and chart recorder compared to master.	1	6	Carry on with current process checks

Hazard Analysis Critical Control Points (HACCP)

Hazard Analysis Critical Control Points (HACCP) is a management system in which product or process safety can be addressed through the analysis and control of biological, chemical, and physical hazards from raw material production to manufacturing, distribution and use of the finished product. HACCP was developed by the food industry.

The HACCP system identifies specific hazards and measures for their control. Examples of hazards within the pharmaceutical setting are: environmental aspects of the facility (environmental conditions, hygiene aspects); material flow; manufacturing steps; personnel hygiene and gowning and technical aspects relating to process design. HACCP is a tool which is used to focus more on prevention and can be used to reduce the reliance upon in-process monitoring or end-product testing. HACCP systems are generally useful for examining changes, such as advances in equipment design, processing procedures or technological developments.

The HACCP system consists of the following seven principles:

- PRINCIPLE 1:   Conduct a hazard analysis.
- PRINCIPLE 2:   Determine the Critical Control Points (CCPs).
- PRINCIPLE 3:   Establish critical limit(s).
- PRINCIPLE 4:   Establish a system to monitor control of the CCP.
- PRINCIPLE 5:   Establish the corrective action to be taken when monitoring indicates that a particular CCP is not under control.
- PRINCIPLE 6:   Establish procedures for verification to confirm that the HACCP system is working effectively.
- PRINCIPLE 7:   Establish documentation concerning all procedures and records appropriate to these principles and their application.

An approach to for conducting a HACCP risk assessment is through twelve steps, which relate to the seven principles outlined above:

1. Assemble HACCP team

As with any risk assessment tool, the team involved in undertaking the risk assessment should be knowledgeable about the process under examination (that is a multidisciplinary team). From the outset the scope of the HACCP plan should be identified. The scope should describe what is to be looked at and the general classes of hazards to be addressed (e.g. microbiological).

2. Describe the product or process

A full description of the product or process should be drawn up, including relevant operational parameters relating to the environment.

3. Identify process stage and subsequent processing

The process stage at which the hazard occurs or could occur should be examined. Consideration should be given to the next stage of processing so that the relative risk can be evaluated (e.g. if the product is filtered at the next stage the risk might be lesser than if the product was placed into an open vessel).

4. Construct flow diagram

The flow diagram should be constructed by the HACCP team. The flow diagram should cover all steps in the operation. When applying HACCP to a given operation, consideration should be given to steps preceding and following the specified operation.

5. Confirmation of flow diagram

The HACCP team should confirm the processing operation against the flow diagram during all stages and hours of operation and amend the flow diagram where appropriate.

6. List all potential hazards associated with each step, conduct a hazard analysis, and consider any measures to control identified hazards

This relates to principle 1 listed above. For this the HACCP team should list all of the hazards that may be reasonably expected to occur at each step from primary production, processing, manufacture, and distribution until the point of consumption.

The HACCP team should next conduct a hazard analysis to identify for the HACCP plan which hazards are of such a nature that their elimination or reduction to acceptable levels is essential to the production of clean product.

In conducting the hazard analysis, the following should be included (where applicable):

- the likely occurrence of hazards and severity of their adverse health effects;
- the qualitative and/or quantitative evaluation of the presence of hazards;
- survival or multiplication of microorganisms of concern;
- production or persistence in foods of toxins, chemicals or physical agents; and,
- conditions leading to the above.

The HACCP team must then consider what control measures, if any, exist which can be applied for each hazard (more than one control measure may be required to control a specific hazard(s) and more than one hazard may be controlled by a specified control measure).

7. Determine Critical Control Points

This connects with the second principle described above. At each step there may be more than one critical control point (CCP) at which control needs to be applied to address the same hazard (this can also include places to monitor, indeed HACCP is quite an effective tool to use when selecting locations for microbiological environmental monitoring). Sometimes decision trees can be useful when deciding which controls are of relevance. If a hazard has been identified and it is decided that control is necessary, and where no control measure exists, a control measure should be adopted.

8. Establish critical limits for each CCP

Critical limits, in connection to the third principle, must be specified and validated for each CCP. In some cases more than one critical limit will be required at a particular step. Criteria often used include measurements of temperature, time, moisture level, pH, water activity. microbial bioburden and endotoxin.

9. Establish a monitoring system for each CCP

Monitoring is the scheduled measurement or observation of a CCP relative to its critical limits (as set out in principle four). The monitoring procedures must be able to detect loss of control at the CCP (for this short term or long term data analysis may be required). Where possible, process adjustments should be made when monitoring results indicate a trend towards loss of control at a CCP. Normally physical and chemical measurements can be undertaken in 'real time' or shortly after a sample is taken, allowing some actions to be considered for risk situations (such as a rise in temperature which might trigger microbial growth). Although rapid microbiological methods are becoming more commonplace, microbiological results are often available sometime after the event has happened.

10. Establish corrective actions

In line with the fifth principle, specific corrective actions must be developed for each CCP in the HACCP system in order to deal with deviations when they occur. The actions must ensure that the CCP has been brought under control. Actions taken must also include a product risk assessment.

11. Establish verification procedures

Establishing procedures for verification forms part of the sixth principle. For this, test results (often as trend reports) and auditing records are particularly useful. In addition, deviation reports and a review of the effectiveness of corrective and preventative actions can provide valuable information.

12. Establish Documentation and Record Keeping

Efficient and accurate record keeping is essential to the application of a HACCP system and is the core of the seventh principle. HACCP procedures should be documented. Documentation examples are:

• Hazard analysis;
• CCP determination;
• Critical limit determination.
• Record examples are:
• CCP monitoring activities;
• Deviations and associated corrective actions;
• Modifications to the HACCP system.

The key points involved in constructing a HACCP are summarized below:

1. Assemble the HACCP team.
2. Describe the product or process.
3. Identify process steps.
4. Construct a flow diagram.
5. Review the diagram, consider any missing steps.
6. List potential hazards.
7. Conduct a hazard analysis.
8. Consider control measures.
9. Identify critical control points.
10. Establish limits for each critical control point.
11. Set up a monitoring system for each control point.
12. Establish corrective actions.
13. Verify findings.
14. Document the outcome.

The advantages of the HACCP approach are that it allows for a systematic overview of the process for the evaluation of each processing step, and allows each step to be examined the possible risks, and allows for the specification of the measures required for controlling each risk. The primary disadvantage is that, unlike FMEA, HACCP cannot be used to rank or prioritize risks. HACCP is also less effective for focusing on an aspect for the process for the objective of HACCP is to map out an entire process.

Risk ranking and filtering

The concept of risk ranking has been touched upon in this chapter. This can be added onto some of the risk assessment tools and can be used to prioritize activities and allows different risk to be compared. Statistical tool like histograms, control charts or Pareto charts can aid and facilitate decision making.

In keeping with this, risks are sometime grouped into:

- primary Risks (Rsp)
- residual Risks (Rsr)

Summary

This chapter has presented an overview of risk management and has focused on some of the primary tools for conducting risk assessment. In doing so the differences between the risk tools have been outlines, together with some of the strengths and weaknesses of each approach.

The chapter has also highlighted that, in undertaking risk assessments, it is important to attempt risk mitigation and to attempt to lower the risk until the risk can be lowered no further. This involves identifying actions to reduce the probability of event and to reduce the severity of event. When this can be taken no further the focus should move towards providing a more reliable detection method designed to initiate a reliable response to a risk event. A further important consideration is that risk assuming actions should be periodically re-assessed.

References and further reading

British Standard BS: IEC61882:2002 Hazard and operability studies (HAZOP studies)- Application Guide British Standards Institution.

Deming WE. (1986): Out of the Crisis. Cambridge, MA: Massachusetts Institute of Technology Center for Advanced Engineering Study.

Ingram, D (2009): Technical Problem Solving. Journal of Validation Technology, Winter 2009, pp 64 – 70

Jackson S. Successfully implementing total quality management tools within healthcare: what are the key actions? Int J Health Care Qual Assur 2001;14(4):157-63.

Kemppainen J.K. (2000): The critical incident technique and nursing care quality research. J Adv Nurs; 32(5):1264-71.

McNally MK, Page MA, Sunderland VB. Failure mode and effects analysis in improving a drug distribution system. Am J Health Syst Pharm 1997;54:17-7.

Phoenix., K. and Andrews, J. (2003). Adopting a Risk - Based Approach to 21 CFR Part 11 Assessments, Pharmaceutical Engineering, July/August2003, Volume 23, Number 4

Tague, N.R. (2004). The Quality Toolbox, Second Edition, ASQ Quality Press, pp236-240.

Tran, N.L., Hasselbalch, B., Morgan, K., Claycamp, G. "Elicitation of Expert Knowledge About Risks Associated with Pharmaceutical Manufacturing Process." Pharmaceutical Engineering, July/August 2005: 24-38

4 CASE STUDY - ENVIRONMENTAL MONITORING RISK ASSESSMENT TOOLS

Introduction

Environmental Monitoring describes the microbiological testing undertaken in order to detect changing trends of microbial counts and micro-flora growth within cleanroom or controlled environments. The results obtained provide information about the physical construction of the room, the performance of the Heating, Ventilation, and Air-Conditioning (HVAC) system, personnel cleanliness, gowning practices, the equipment, and cleaning operations.

Over the past decade, environmental monitoring has become more sophisticated in moving from random sampling, using an imaginary grid over the room and testing in each grid, to the current focus on risk assessment and the use of risk assessment tools to determine the most appropriate methods for environmental monitoring.

This chapter explores current trends in the application of risk assessment to the practice of environmental monitoring by examining the following key areas:

- **Determining the Frequency of Monitoring**: Using the concept of risk assessment to decide how often to monitor different types of cleanrooms
- **Risk Assessment Tools:** Applying risk assessment tools to establish methods for environmental monitoring
- **Numerical Approaches:** Considering a numerical approach to assess risk data using a case study of an aseptic filling operation

The examples used are from a sterile drug manufacturing facility and focus mostly on aseptic filling; however, the concepts and tools are applicable to the environmental monitoring of other types of manufacturing and packaging operations.

Determining the frequency of monitoring

In developing an adequate environmental monitoring programme, there should be a balance between using resources efficiently and monitoring at sufficiently frequent intervals so that a meaningful picture can be obtained. Sources of guidance with respect to monitoring frequencies are very limited within Europe, and the monitoring frequencies specified within the United States Pharmacopoeia (USP) <1116> may not be suitable for all facilities. Some guidance can be obtained from the International Organization for Standardization's (ISO) standards: principally ISO 14644 and ISO 14698. However, these do not always fit with regulatory guidance documents because they apply to controlled environments across a range of industries other than pharmaceuticals, where standards can be higher (Jahnke, 2001).

When establishing an environmental control programme, the frequency of monitoring different controlled areas can be determined based on 'criticality factors' relevant to each specific area.

Criticality Factors

The establishment of a criticality scheme on which to base monitoring frequencies is designed to target monitoring of critical process steps. Therefore, the final formulation process would receive more monitoring than an early manufacturing stage with a relatively closed process.

Using a criticality factor is a means of assigning a monitoring frequency based on the risk assessment of each critical area. The risk assessment relates to the potential product impact from any risk. For example, an area of open processing at an ambient temperature, a long exposure time, and the presence of water, would constitute a high risk and would attract a higher risk rating. In contrast, an area of closed processing, in a cold area, would carry a substantially lower risk and associated risk rating.

Using a range of 1 to 6, with 1 being the most critical and 6 the least critical, a score of 1 would be assigned to an aseptic filling operation; a score of 2 to final formulation, a score of 3 to open processing, and so on. Each user must adapt such a scheme to his or her particular area and defend it by way of a supportable rationale. An example of monitoring frequencies under such a scheme can be seen in *Figure 1*, and an example of its application is seen in *Figure 2*:

Figure 1

Criticality Factors of Monitoring Frequencies

Criticality Factor	Frequency of Monitoring
1	Daily or Each Batch
2	Weekly
3	Fortnightly or Bi-weekly
4	Monthly
5	Three-monthly or Quarterly
6	Six-monthly or Semi-annually

Figure 2: Application of Criticality Factors

Environmental Criticality Factor	Likelihood of Environmental Impact on Finished Product	Definition	Monitoring Frequency
1	Highly Likely	Aseptic filling where no further processing takes place. Here the risk of contamination would have a considerable product impact because contaminants could not be reduced or removed by further processing.	Daily or Each Batch
2	Likely	Area of final formulation. This may apply to an area where the final process is a sterilizing grade filter.	Weekly
3	Moderately Likely	Direct or indirect exposure of the product to the environment is somewhat likely to introduce	Fortnightly or Bi-Weekly

Environmental Criticality Factor	Likelihood of Environmental Impact on Finished Product	Definition	Monitoring Frequency
		contaminants. This may also apply to an area that is at ambient temperature and where there is a high water presence.	
4	Unlikely	This may apply to cold areas where little or no open processing takes place.	Monthly
5	Very Unlikely	Indirect exposure to the environment is highly unlikely to introduce contaminates that could affect the finished product. If a contaminant were to be introduced, sufficient downstream controls and/or the use of preservative agents are highly likely to remove	Every 3Months

Environmental Criticality Factor	Likelihood of Environmental Impact on Finished Product	Definition	Monitoring Frequency
		and significantly reduce contaminates.	
6	High Unlikely	An area that is uncontrolled or where microbial contamination is very unlikely, such as a freezer	Every 6 months

Each controlled area would be evaluated against set criteria and, with the use of a series of guiding questions, the monitoring frequency would be determined. Decision criteria include considerations in two category areas: areas of higher weighting and areas of higher monitoring frequency. Examples of these categories follow:

➤ *Giving Higher Weighting to -*

- ✓ 'Dirtier' activity performed in a room adjacent to a clean activity, even if the clean activity represents later processing
- ✓ Areas that have a higher level of personnel transit (given that people are the main microbiological contamination source). This may include corridors and changing rooms.
- ✓ Routes of transfer
- ✓ Areas that receive in-coming goods
- ✓ Component preparation activities and sites
- ✓ Duration of activity (such as a lower criticality for a 30-minute process compared to a six-hour operation)

➤ *Having Higher Monitoring Frequencies for -*

- ✓ Warm or ambient areas as opposed to cold rooms
- ✓ Areas with water or sinks as opposed to dry, ambient areas
- ✓ Open processing or open plant assembly compared to processing that is open momentarily or to closed processing (where product risk exposure time is examined)
- ✓ Final formulation, purification, secondary packaging, product filling, etc.

Once the monitoring frequency for each controlled area is determined, it should be reviewed at regular intervals. This review may invoke changes to a room's status, and hence, its monitoring frequency, or to changes for different sample types within the room. For example, it may be that after reviewing data for one year, surface samples produce higher results than air samples for a series of rooms. In this event, the microbiologist may opt to vary the frequency of monitoring and take surface samples more often than air samples. There would also be an increased focus on cleaning and disinfection practices, and their frequencies, based on such data (Sandle, 2004b).

When both types of monitoring are producing low level counts, the balance of risk would be towards air samples. This is because air samples are direct indicators of the quality of the process and assign a level of control to the process, whereas surface samples are indicators of cleaning and disinfection. If the results of surface samples are generally satisfactory, as indicated by trend analysis, then either the number of samples or the frequency at which they are taken can be reduced. If subsequent data showed an increase in counts, the monitoring frequency could easily be restored. Indeed, all types of monitoring frequencies may increase as part of an investigation, as appropriate. Therefore, the criticality factor approach not only sets the requirement for a room, it can also be used to vary the sample types within a room (Ljungqvist and Reinmuller, 1996).

Risk assessment tool

Once the status for each room has been selected, a risk assessment procedure is required to determine locations for environmental monitoring. Such risk-based approaches are recommended in ISO 14698 and regulatory authorities are increasingly asking drug manufacturers about this subject.

Risk-based approaches include Failure Modes and Effects Analysis (FMEA), Fault Tree Analysis (FTA), and Hazard Analysis Critical Control Points (HACCP), all of which employ a scoring approach. (Other approaches include: Failure Modes, Effects and Criticality Analysis (FMECA), Hazard Operability Analysis (HAZOP), Quantitative Microbiological Risk Assessment (QMRA), Modular Process Risk Model (MPRM), System Risk Analysis (SRA), Method for Limitation of Risks, and Risk Profiling.) At present, no definitive method exists, and the various approaches differ in their process and in the degree of complexity involved. However, the two most commonly used methods appear to be HACCP, which originated in the food industry, and FMEA, which was developed for the engineering industry (Whyte and Eaton, 2004a).

These various analytical tools are similar in that they involve:

- Constructing diagrams of work flows.
- Pin-pointing areas of greatest risk.
- Examining potential sources of contamination.
- Deciding on the most appropriate sample methods.
- Helping to establish alert and action levels.
- Taking into account changes to the work process and seasonal activities.

These risk assessment approaches are not only concerned with selecting environmental monitoring locations. They integrate the environmental monitoring system with a complete review of operations within the cleanroom to ensure those facilities, operations, and practices are also satisfactory. The approaches recognise a risk, rate the level of the risk, and then set out a plan to minimise, control, and monitor the risk. The monitoring of the risk will help to determine the frequency, locations for and level of environmental monitoring.

This chapter explores an example from three different techniques:

- A simple conceptualisation of risk using a table
- HACCP
- FMEA

Tabular Approach

An example using a simple table for analyzing risk in environmental monitoring situations appears in *Figure 3*.

Figure 3

Tabular Approach to Risk Assessment

Area or Equipment: Sterility Testing Isolator		
Risk: Contamination due to build-up of microbial counts in the isolator environment		
Failure or Situation: Failure to adequately clean after use		
Effect	**Minimising the Risk (Mitigations to Reduce Risk)**	**Monitoring**
• When isolators are not cleaned regularly, there is a possibility of micro-organisms remaining in the environment.	• Cleaning surfaces using water to remove dirt or spillages prior to the application of a suitable disinfectant. • The disinfectant used must have a wide spectrum of efficacy, but not be aggressive to the isolator material. • The isolator should be designed so that it is easy to clean.	• An environmental monitoring programme (using settle plates, air samples, contact plates, swabs, or finger plates) will show the areas of greatest risk. This data should be examined for trends. • For out-of-limits environmental monitoring results, appropriate Corrective and Preventive Actions (CAPA) should be put in place.

HACCP

➤ **The seven principles behind constructing an HACCP analysis consist of:**

1. Identifying hazards or contamination risks and assessing their severity
2. Determining Critical Control Points (CCPs)
3. Establishing critical limits
4. Establishing a system to monitor and control CCPs
5. Establishing corrective action when a CCP is not under control
6. Establishing procedures for verification to confirm that the HACCP system is working effectively
7. Establishing documentation and reporting systems for all procedures

Each of these seven key points is a vital step in developing the risk assessment.

➤ **The seven points include:**

1. Construct a risk diagram, or diagrams, to identify sources of contamination. Diagrams should show sources and routes of contamination.

   Examples include:

   - ✓ Areas adjacent to cleanroom or Isolator (e.g.: airlocks, changing rooms)
   - ✓ Air supply and Room air
   - ✓ Surfaces
   - ✓ People
   - ✓ Machines and Equipment

2. Assess the importance of these sources and determine whether or not they are hazards that should be controlled.

Examples include:

- ✓ Amounts of contamination on, or in, the source that is available for transfer
- ✓ Ease by which the contamination is dispersed or transferred
- ✓ Proximity of the source to the critical point where the product is exposed
- ✓ Ease with which the contamination can pass through the control method

The use of a scoring method can be very helpful in assessing the relative importance of these contamination sources.

3. Identify the methods that can be used to control these hazards.

For example:

- ✓ Air Supply: High Efficiency Particulate Air (HEPA) filters
- ✓ Dirty Areas adjacent to Cleanroom or Isolator: differential pressures, airflow movement
- ✓ Room Air: air change rates, use of barriers
- ✓ Surfaces: sterilisation, effectiveness of cleaning and disinfection procedures
- ✓ People: cleanroom clothing and gloves, room ventilation, training
- ✓ Machines and Equipment: sterilisation, effectiveness of cleaning, exhaust systems

4. Determine valid sampling methods to monitor either the hazards or their control methods or both.

For example:

- ✓ HEPA filter integrity tests
- ✓ Air supply velocity, air change rates
- ✓ Room pressure differentials
- ✓ Particle counts
- ✓ Air samplers, settle plates, contact plates, etc.

5. Establish a monitoring schedule with 'alert' and 'action' levels and corrective measures to be taken when these levels are exceeded.

    For example:

    ✓ The greater the hazard, the greater the amount of monitoring required
    ✓ Trend analysis for alert and action levels, in or out of control

6. Verify that the contamination control system is working effectively by reviewing key targets like product rejection rate, sampling results, control methods, and so on. These may require modification over time.

    For example:

    ✓ System for data review
    ✓ Examine filling trials
    ✓ Audits
    ✓ Reassess - hazards, effectiveness of control systems, frequency of monitoring, appropriateness of alert and action levels

7. Establish and maintain documentation.

    For example:

    ✓ Describe the steps being taken
    ✓ Describe the monitoring procedures
    ✓ Describe the reporting and review procedures

Before implementing HACCP, it is important to train all staff involved in the process and to use a multi-disciplinary team. For example, the team may be made up of personnel from Production, Engineering, Quality Control (QC), Quality Assurance (QA), Validation, and so on.

FMEA

FMEA schemes vary in their approach, scoring, and categorisation. All methods share a numerical approach. The example presented here, based on a sterility testing isolator, assigns a score (from 1 to 5) to each of the following categories:

➤ *Severity*
➤ *Occurrence*
➤ *Detection*

Where:

✓ Severity is the consequence of a failure
✓ Occurrence is the likelihood of the failure happening based on past experience
✓ Detection is based on the monitoring systems in place and on how likely a failure can be detected.

By asking a series of questions, each main part of the cleanroom or isolator system can be grouped or classified into key parts.

Such questions include:

✓ What is the function of the equipment? What are its performance requirements?
✓ How can it fail to fulfil these functions?
✓ What can cause each failure?
✓ What happens when each failure occurs?
✓ How much does each failure matter? What are its consequences?
✓ What can be done to predict or prevent each failure?
✓ What should be done if a suitable proactive task cannot be found?

The scoring is 1 (very good) to 5 (very bad). Therefore, a likelihood of high severity would be rated 5; high occurrence rated 5; but a good detection system would be rated 1.

Using these criteria, a final FMEA score is produced from:

Severity score x Occurrence score x Detection score

Decisions on further action will depend upon the score produced. There is no published guidance on what the score that dictates some form of action should be. However,. A suggested score is 27 for the cut-off value where action was required. This is based on 27 being the score derived when the mid-score is applied to all three categories (i.e., the numerical value '3' for severity 3 x occurrence 3 x detection 3) and the supposition that if the mid-rating (or a higher number) is scored for all three categories, then at a minimum, the system should be examined in greater detail.

Figure 4 Isolator Operation Example

An example of one area of an isolator operation, and the risks associated with the room in which the isolator is housed, is examined below.

Description of the Critical Area. The isolator is situated in an unclassified room. There is no requirement to place a sterility testing isolator in a classified room.

FMEA Schematic.

Process Step	Failure Mode	Significance of Failure	Severity of Consequence (score)
Loading isolators pre-sanitisation, performing sterility testing	That contamination from the room could enter transfer or main isolators	Reduced efficiency of transfer isolator sanitisation, contamination inside main isolator	3

Measures to Detect Failure	Occurrence (score)	Detection Systems	Detection (score)
Would be shown from reduced evaporation rate for isolator sanitisation, poor environmental monitoring results in main isolator, potential sterility test failures,. Sanitisation cycle has been validated using biological indictors of 10^6 spores.	1	Isolator room is monitored monthly for viables and particles, staff-wear over-shoes on entry, Dycem mat in place, entry to room has controlled access, environmental monitoring performed inside main isolator. Isolators are at positive pressure to the room and air is HEPA filtered.	1

FMEA score: $3 \times 1 \times 1 = 3$

Analysis: There is no problem considered from the room environment. Entry to the room is controlled; the sanitisation cycle has been challenged with a level of micro-organisms far greater than would ever be found in the environment (spores of *Geobacillus stearothermophilus*); all items entering the isolator are sanitised (using a chlorine dioxide-based sporicidal disinfectant); and the isolator itself is an effective, positive pressure barrier to the outside (at >15 Pascals).

As detailed earlier, environmental monitoring is performed inside the isolator during testing. This monitoring, which has an action level of 1 CFU (Colony Forming Unit), is designed to detect any potential contamination inside the isolator environment.

Numerical approaches

A third component of the risk assessment approach is to evaluate that a risk once an activity has taken place. Then, by using a largely numerically-driven set of tools, repeatability and reproducibility can be ensured. Examples of individual out-of-limits results and data-sets relating to an operation are examined below, using examples from an aseptic filling process. Following this an example of an overall assessment of different processes over time is explored. Numerical approaches are useful in applying a level of consistency between one decision and another.

Individual Assessments

The section below details some methods that can be used to quantify the risk of contamination in pharmaceutical cleanrooms. The models outlined are based on the work performed by Whyte and Eaton (2003a and b).

➤ *Estimating the Risk to Product Using Settle Plate Counts*

The method applies to the assessment of settle plates at the point-of-fill, under the Grade A zone. It allows an estimate of the probable contamination rate to the product as derived from the following equation:

Contamination rate (%) =

Plate count x **Area of product** x **Time product exposed** x 100
                 Area of Petri dish   **Time settle plates exposed**

The fixed value is the area of the Petri dish, which for a 90mm plate, is 64 cm^2.

Figure 5

Micro-Organism Risk Categories

Percentage	Risk
<0.03%	Low
>0.03 – 0.09%	Medium
≥0.1%	High

➤ *Finger Plate Assessment*

The formula can readily be applied to operations that relate to Grade A operations, for example: filtration connection, vessel to filling machine connection, the filling activity, and loading a freeze-dryer. Where the operator is only present in the Grade B room and has no impact on the Grade A operation, this is automatically considered to be low risk if there are no other special factors. (Low risk does not imply lack of action or assessment. However, it aims conceptualisation of the result in terms of probable risk to the batch.)

The following formula can be used:

Microbial count x Location x Method of intervention x Duration of operation

Where:

Microbial Count	=	Count in cfu for the plate	
Location	=	Area of the filling machine, or other location to which the plate relates	In this
Activity	=	Whether the hand directly touched part of the filling machine or if utensils were used	
Duration	=	Length of the activity in seconds	

77

example of a finger plate assessment, the location, activities, and duration require weighting. Examples of logic that applies to the rating of the location, activities, and duration categories can be seen in *Figures 6, 7, and 8*, respectively.

Figure 6

Weighting Location Example

Location	Rating*	Reason
General part of machine not close to filling zone	0.5	Data from air-flow patterns suggests very low risk of contamination movement into the unidirectional air-stream over the filling zone
Off-load	0.5	Off-load areas are present for all filling machines. The bottles and vials are partially stoppered and utensils are normally used. The likelihood of contamination is considered to be low.
On-load	1.0	On-load areas are present for all filling machines. The bottles and vials are not stoppered, although utensils are normally used. The likelihood of contamination is higher than for off-load.
Stopper bowl	1.5	Stopper bowls are present for filling machines. A direct intervention into the bowl could result in micro-organisms being deposited onto stoppers. The risk of this is considered higher than the risk with on-load or off-load activities, although such an intervention is rare.
Freeze-dryer loading	1.5	This is a direct intervention Grade A activity. However, vials and bottles are partially stoppered and are contained with cassettes.
Point-of-fill: air-sample placement	2.0	The placement of an air-sampler does not involve the touching of any filling equipment (such as needles, balances etc.). However, as a direct intervention into the Grade A zone it is a higher risk than those parts of the filling machine previously examined.

Location	Rating*	Reason
Filtration transfer	2.0	The connection of a vessel for the purpose of transferring a product into the Aseptic Filling Suite requires human intervention and aseptic technique. If this process becomes contaminated, this could affect the product. The time taken to perform the connection is normally very short (under 30 seconds), which reduces the risk.
Machine connection	2.5	The connection of a vessel to the filling machine requires human intervention and aseptic technique. If the transfer line is contaminated, this could cause contamination to the product.
Point-of-fill: intervention	2.5	A direct intervention, where for example, filling needles are re-adjusted, is the highest risk rating. Counts associated with such activities require detailed examination.

*Where 0.5 represents the lowest risk and 2.5 the highest risk.

Figure 7

Weighting Activities Example

Activity Method	Rating	Reason
Using forceps	0.5	The operative does not directly touch the machine and the utensils used are sterile.
Hand	1.0	The operative directly touches the machine, thereby creating a greater risk. However, it is procedure to sanitise hands prior to undertaking the operation.

Figure 8

Weighting Duration Example

Duration	Rating	Reason
Less than 30 seconds	0.5	The length of the intervention is considered minimal.
30 seconds – 120 seconds	1.0	The intervention is at the average* time taken.
Plus 120 seconds	1.5	The intervention has taken longer than average*.

* Based on the average time taken for media simulation trials, based on data from a UK pharmaceutical facility.

➤ *Finger Plate Assessment Worked Example*

A finger plate with a count of 1 cfu for an activity at point-of-fill, using forceps, that lasts for one minute.

Microbial count x Location x Method of intervention x Duration of operation

1 x 2.5 x 0.5 x 1   =   1.25

The score produced would be rated according to standard risk assessment categories:

Score	Risk
1 – 3	Low
4 – 8	Medium
9+	High

These risk ratings are based, in part, on the worked example. Based on historical data over the past six-months, the highest record example of a Grade A intervention finger plate is a count of 2 cfu: using forceps to retrieve a fallen vial and lasting for more than 120 seconds. This would have given a score of 7.5, which falls within the medium risk category. The user should develop a scheme that fits his or her facility (Whyte and Eaton, 2004b).

➤ *Surface Sample Assessment*

The following formula can be applied to filling and filtration activities.

Microbial count x Risk Factor A x Risk Factor B x Risk Factor C

Where:

Risk Factor A = Proximity to critical area
Risk Factor B = Ease of dispersion of micro organisms
Risk Factor C = Effectiveness of control measure

Samples are taken using contact plates and swabs and are all post-operation.

✓ The following approach can be used in setting the risk factors:

» The first step is to assign the risk (A) factor based on proximity of location to the critical area (filled product). The logic illustrated in *Figure 9* may be used to determine risk factor A.

Figure 9

Determining Risk Factor A

Manufacturing Stage/ Location	Risk Factor (A)	Reason
Filtration room product contact	2.0	Samples of the transfer line. May indicate potential for contamination to affect product.
General filling room or filtration room area (Grade B)	0.5	The samples reflect room cleanliness and general trends only. The impact upon the Grade A activity is low – *unless* the same micro-organism has been detected from a Grade B action level sample. In this event, the risk factor increases to 1.
Machine general (non-product contact)	1.0	Samples indicate state of general machine cleanliness but the risk of exposure of product to contaminant is low - *unless* the same micro-organism has been detected from a Grade B action level sample. In this event, the risk factor increases to 2.5.
Machine product contact site	2.5	Sites include utensils, filling needles, and stopper bowls. Direct contact with product: highest risk.

» The second step is to assign a risk factor (B) based on ease of dispersion or transfer of micro-organisms. See *Figure 10* for an example of the reasoning that might support risk factor B.

Figure 10

Determining Risk Factor B

Manufacturing Stage/ Location	Risk Factor (B)	Reason
Filtration room product contact	0.5	Connection is a short activity (less than thirty seconds) performed under Grade A UDAF[1] protection; operator wears sanitised gloves.
General filling room or filtration room area (Grade B)	1.0	Periphery to the Grade A zone. Risk of transfer is low. Risk rating would increase to 1.5 if a Grade A and a Grade B sample exceeded action level and was characterised as the same microbial species.
Machine general (non-product contact)	1.5	Location is within the critical area, but not directly in product contact area. Some risk of transfer exists, but protective measures should prevent this.
Machine product contact site	2.5	Sites include utensils, filling needles, and stopper bowls that are directly within the critical zone. Direct contact with product: highest risk.

» The third step is to weight the risk factor 'C' by assessing the effectiveness of the control measure. See *Figure 11* for an example of this assessment.

---

[1] Unidirectional Air Flow

Figure 11

Determining Risk Factor C

Manufacturing Stage/ Location	Risk Factor (C)	Reason
Filtration room product contact	0.5	Grade A UDAF and sterilised components; operator wears two pairs of gloves and sanitises hands.
General filling room or filtration room area (Grade B)	0.5	Floor is sanitised. Barrier exists by way of filling machine doors and UDAF.
Machine general (non-product contact)	1.0	Sterilised machine components; lines are wiped with disinfectants; UDAF protection.
Machine product contact site	1.5	Sterilised machine components; no direct intervention; UDAF protection; however, site is in direct contact with product.

➤ *Surface Sample Worked Example:*

Where a count of 2 is detected from a conveyor belt (a filling machine non-product contact location)

Using the formula:

Microbial count x Risk Factor A x Risk Factor B x Risk Factor C

$$2 \ \times \ 1 \ \times \ 1.5 \ \times \ 1 \ = \ 3$$

Risks can be scored against standard risk assessment categories:

Score	Risk
1 – 3	Low
4 – 8	Medium
9+	High

This scoring scheme is based on contamination of a product contact site being high risk by virtue of its direct proximity to the critical area or the product.

A count of 1 cfu on one of these product contact site locations would give a score of 9.4. In most filling zones and clean zones, sample results from product contact sites would be expected to record zero counts for 999 samples out of every 1000. Whereas, a count of 3 from a non-product contact site would result in a medium risk category.

➤ *Air-sample Assessment*

Approaches are available for the risk assessment of active air samples that use a numerical system. However, the formulae associated with these are difficult to calculate in practice because often all information is not available and assessment of variables, such as impaction speed, are not readily calculable. Therefore, a qualitative assessment, such as the one illustrated below, may be more suitable.

An example of the numerical approach:

Airborne microbial count (cfu /m³) x deposition velocity of micro-organisms from air (cm/s) x area of product exposed (cm²) x time of exposure (s)

Alternatively, non-numerical risk assessment can be used based on the proximity and the operation. See *Figure 12* for the example.

Figure 12

Non-Numerical Risk Assessment Example

Activity, Area, Proximity	Risk	Reason
General room environment during filling or filtration, away from Grade A zone	Low	Provided there are no Grade A interventions and there are no counts recorded at Grade A. Where counts are recorded at Grade A and there is a micro-organism match, then the category is re-defined as medium risk.
Grade A, near critical zone	Medium	Close, but not at the point-of-fill
Grade A point-of-fill at critical zone	High	At point-of-fill.

Note: *Where growth is detected on the operator who placed the air-sampler at Grade A, and this is shown to be the same micro-organism, the category of risk is increased by one (i.e.: low becomes medium and medium becomes high).*

➤ *Assigning a Risk Factor to Areas of the Filling Room*

The location where a high bio-burden is isolated within the filling area is arguably of greater consequence than the actual count. The location can be given a risk rating in relation to its proximity to the critical zone, ease of dispersion or transfer, and effectiveness of control methods.

The table shown in *Figure 13* is proposed as a tool for risk assessment and to aid investigations. It supplements the risk assessment tools that have been previously examined.

Figure 13

Filling Room Risk Assessment Example

Ease of Dispersion or Transfer of Micro-organisms	Proximity or Location of Source from Critical Area	Effectiveness of Control Method
None	None	None
Very low, e.g.: fixed place on sterilised area	Low, e.g.: at extreme limit of room away from filling zone	Low, e.g.: barrier control; UDAF
Medium, e.g.: product contact device	Medium, e.g.: general area of cleanroom near filling machine or at edges of Grade A zone	Medium, e.g.: sanitisation
High, e.g.: gloved hands of operators with direct contact with product	High, e.g.: within the critical area	High, e.g.: no effective controls

The type of product and whether further processing occurs can also influence the risk factors. See *Figure 14* for an example. In considering batches with a high risk rating, further processing of the product can be considered and ranked (1 = lowest risk, 4 = highest risk):

Figure 14

Product-related Risk Assessment Factors

Product	Rank	Reason
Freeze-dried	1	Freeze-drying will destroy most micro-organisms
Liquid product, heat treated	2	Undergoes pasteurisation – effective against most non-spore forming micro-organisms
Intra-muscular product	3	Small volume; intra-muscular route
Intravenous product, no further treatment	4	Intravenous route; no further processing

An Overall Assessment

The approach taken for an overall assessment involves the historical examination of a number of operations and assigning a value above which the operation is considered to be atypical. A 95% cut-off was considered to be the most suitable cut-off point.

Conclusion

The use of risk assessment approaches is an important current Good Manufacturing Practice (cGMP) topic in microbiological environmental monitoring. This chapter has outlined some possible tools for such a risk assessment approach; however, each suite of cleanrooms or isolator will be subtly different. The microbiologist must consider each aspect of the environment and decide what level of monitoring best suits his or her system, and then must justify the techniques used and the locations selected.

The approach adopted should be detailed in a written rationale and approved by senior management. After this, a rigorous and defensible system will be in place to satisfy regulatory expectations, and to aid the user in assessing the risk of problematic environmental monitoring situations or results.

References

Anon. (1998), The Gold Sheet, Vol. 32, No.10, October 1998

BS EN ISO 14698 – 1:2003: 'Cleanrooms and associated controlled environments – Biocontamination control – Part 1: General principles and methods'

Code of Federal Regulations, 1998, Title 21, Part 210, Current Good Manufacturing Practice in Manufacturing, Processing, Packaging, or Holding of Drugs – General, 210:3

De Abreu, C., Pinto, T. and Oliveira, D. (2004): 'Environmental Monitoring: A Correlation Study Between Viable and Nonviable Particles in Cleanrooms', Journal of Pharmaceutical Science and Technology, Vol. 58, No.1, January-February 2004, pp45-53

'Guidelines on Sterile Drug Products Produced by Aseptic Processing', FDA, 1987 (draft produced for review in August 2003)

ISO 14644-1 Cleanrooms and Associated Controlled Environments – Classification of Air Cleanliness

ISO 14644-2 Cleanrooms and Associated Controlled Environments – Specifications for Testing and Monitoring to prove continued compliance with ISO 14644-1

ISO 14698-1 Cleanrooms and Associated Controlled Environments Biocontamination Control – Part 1: General Principles

Jahnke, M. (2001): 'Introduction to Environmental Monitoring in Pharmaceutical Areas', PDA

Kaye, S. (1986): 'Efficiency of Biotest RCS as a Sampler of Airborne Bacteria', Journal of Parenteral Science and Technology, Vol. 42, No.5, September-October, pp147-152

Ljungqvist, B. and Reinmuller, B. (1996): 'Some observations on Environmental Monitoring of Cleanrooms', European Journal of Parenteral Science, 1996, 1: 9 –13

Meir, R. and Zingre, H. (2000): 'Qualification of air-sampler systems: MAS-100', Swiss Pharma, 22(2000); pp15 – 21

Ohresser, S., Griveau, S. and Schann, C. (2004): 'Validation of Microbial Recovery from Hydrogen Peroxide-Sterilised Air', Journal of Pharmaceutical Science and Technology, Vol. 58, No.2, March-April 2004, pp75-80

PDA Technical Report No. 13 (revised): 'Fundamentals of an Environmental Monitoring Programme', September / October 2001

PhRMA Environmental Monitoring Work Group (1997): 'Microbiological Monitoring of Environmental Conditions for Nonsterile Pharmaceutical Manufacturing', Pharm. Technol., March, pp58-74

Reich, et al. (2003): 'Developing a Viable Microbiological Environmental Monitoring Program for Nonsterile Pharmaceutical Operations', Pharm. Technol., March, pp92-100

'Rules and Guidance for Pharmaceutical Manufacturers and Distributors' ('EU GMP Guide'), MHRA, 2002

Sandle, T.: 'Environmental Monitoring in a Sterility Testing Isolator', PharMIG News No.1, March 2000

Sandle, T.: 'Microbiological Culture Media: Designing a Testing Scheme', PharMIG News No.2, August 2000

Sandle, T. (2003): 'The use of a risk assessment in the pharmaceutical industry – the application of FMEA to a sterility testing isolator: a case study', European Journal of Parenteral and Pharmaceutical Sciences, 2003; 8(2): 43-49

Sandle, T (2003)[2].: 'Selection and use of cleaning and disinfection agents in pharmaceutical manufacturing' in Hodges, N and Hanlon,

G. (2003): 'Industrial Pharmaceutical Microbiology Standards and Controls', Euromed Communications, England

Sandle, T. (2004): 'General Considerations for the Risk Assessment of Isolators used for Aseptic Processes', *Pharmaceutical Manufacturing and Packaging Sourcer*, Samedan Ltd, Winter 2004, pp43-47

USPNF#25 <1116>

Whyte, W. (2001): 'Cleanroom Technology: Fundamentals of Design, Testing and Operation'

Whyte, W. and Eaton, T. (2004): 'Microbiological contamination models for use in risk assessment during pharmaceutical production', *European Journal of Parenteral and Pharmaceutical Sciences*, Vol. 9, No.1, pp11-15

Whyte, W. and Eaton, T. (2004): 'Microbiological risk assessment in pharmaceutical cleanrooms', *European Journal of Parenteral and Pharmaceutical Sciences*, Vol. 9, No.1, pp16-23

Whyte, W. and Eaton, T. (2004): 'Assessing microbial risk to patients from aseptically manufactured pharmaceuticals', *European Journal of Parenteral and Pharmaceutical Sciences*, Vol. 9, No.3, pp71-79

5 CASE STUDY - ASEPTIC PROCESSING RISK ASSESSMENT (HACCP)

Introduction

Within pharmaceutical manufacturing a shift has taken place from the amassing of test data to a greater emphasis on examining the level of risk to a process and then taking steps to eliminate that risk. This article demonstrates an example of the risk assessment approach by using a case study of transferring a set of sterilised stoppers from an autoclave to a filling machine within a sterile manufacturing facility.

The case study is imaginary, although this is of not of great importance. The exercise is designed to show the advantages of the risk assessment approach on a simple scale. The risk assessment approach adopted is a form of HACCP (Hazard Analysis Critical Control Points).

Approaches to risk assessment

There are various approaches to risk assessment being used in the pharmaceutical industry. Each has their respective merits. The use of risk assessment is recommended in ISO 14698 and regulatory authorities are increasingly expecting to see evidence of risk assessments.

The common approaches include FMEA (Failure Mode and Effects Analysis); FTA (Fault Tree Analysis) and HACCP (Hazard Analysis Critical Control Points), most of which employ a scoring approach . At present, no definitive method exists and the various approaches differ in their process and degree of complexity involved. However, the two most commonly used appear to be HACCP (which originated in the food industry) and FMEA (which was developed for the engineering industry).

These various analytical tools are similar, in that they involve:

- Constructing diagrams of work flows
- Pin-pointing areas of greatest risk
- Examining potential sources of contamination
- Deciding on the most appropriate sample methods
- Helping to establish alert and action levels
- Taking into account changes to the work process / seasonal activities

The risk assessment approaches can allow for a complete review of operations within the clean room to ensure those facilities, operations and practices are also satisfactory. The approaches recognise a risk, rate the level of the risk and then set out a plan to minimise, control and monitor the risk. The monitoring of the risk will help to determine the frequency, locations for and level of environmental monitoring (for example, refer to an article by Sandle [2003]).

This case study uses a modification of the HACCP method which is sometimes called the 'Lifecycle Approach'. The approach adopted is similar to that contained in "Pharmaceutical cGMPs for the 21st Century: A Risk-Based Approach". This approach recommends that the following factors are noted when undertaking HACCP analysis in pharmaceutical manufacturing:

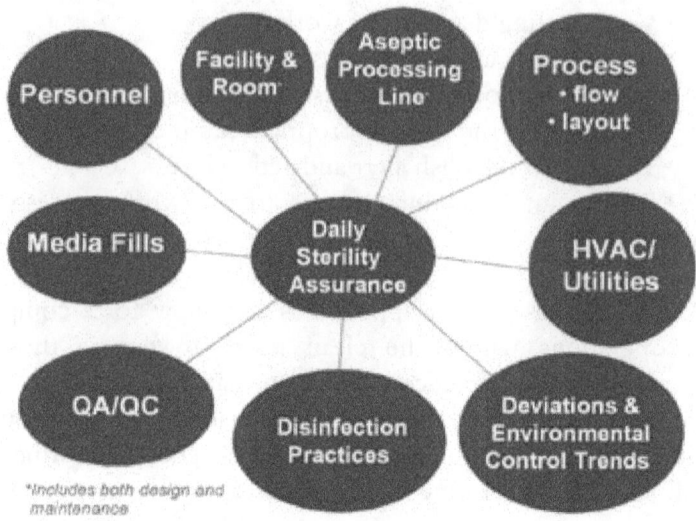

Personnel

Facility & Room

Aseptic Processing Line·

Process
· flow
· layout

Media Fills

Daily Sterility Assurance

HVAC/ Utilities

QA/QC

Disinfection Practices

Deviations & Environmental Control Trends

*Includes both design and maintenance

HACCP was developed during the 1960s and it grew out of a collaboration between NASA, a food company (the Pillsbury Company) and the US Army Natick Laboratories. The objective was to provide a zero-defect food supply for the astronauts. HACCP was derived from Failure Mode Analysis (Gavin and Weddig, 1995). There are two key components of HACCP:

Hazard Analysis: Determining what microbiological, physical, or chemical risks are associated with a process.

Critical Control Point: A point, step, or procedure at which control can be applied.

In general HACCP involves an assessment of the following conditions. These are known in much of the food industry literature as 'the seven pillars':

1. Conducting a hazard analysis List all potential hazards associated with each step, conduct a hazard analysis, and consider any measures to control identified hazards.
2. Determining the Critical Control Points (CCPs).
3. Establishing critical limit(s) Establish critical limits for each

CCP.
4.　　Establishing a system to monitor control of the CCP.
5.　Establishing the corrective action to be taken when monitoring 　　　indicates that a particular CCP is not under control.
6.　　Establishing procedures for verification to confirm that the HACCP system is working effectively.
7.　　Establishing documentation and record keeping.

The Case Study

In the 'imaginary' facility the autoclave that is connected to the filling room is unavailable due to a long term malfunction. The autoclave is used to prepare the stopper for a Parenteral product and the need to process an important batch exists. The production management at the facility have decided that stoppers will be sterilised using a different autoclave within the facility and that the sterilised stoppers will be transported to the filling machine. The stopper swill be loaded, by trained operators, into a sterilised hopper drum and wheeled through part of the facility.

Putting aside any doubts to the suitability of this process the microbiologist has been asked to risk assess the proposed procedure.

In order to do this the microbiologist divided the risk assessment into four steps:

- A route map (where the facility is drawn and the route indicated)
- Identification of hazards (which can be divided into biological; physical; equipment; transport and chemical). This will allow an assessment of existing control measures.
- Process flow.
- Assessment of environmental monitoring. This will determine if the activity is safe to proceed.

Each of these is examined in turn.

Deconstructing the Process

Route map

The first step is to outline a route map. This will focus on transporting the stoppers by the shortest and safest route possible.

Route Map (arrows indicate stopper transfer route from autoclaves to filling room):

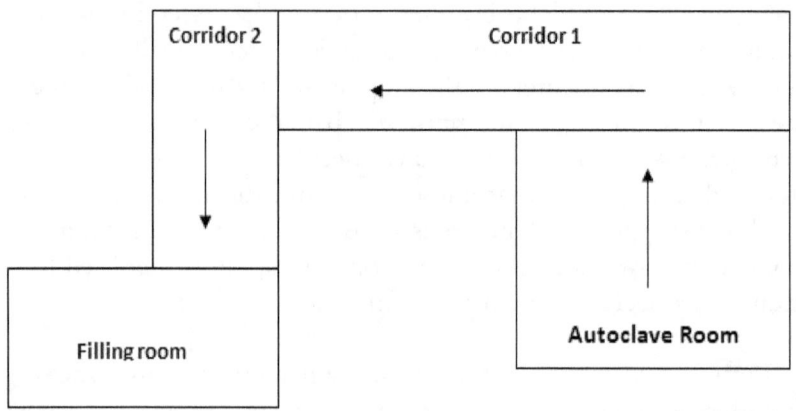

Identification of hazards

The second step is to understand the hazards and the process.

Identification of Hazards

Identification of Hazards

Key Hazard	Source of hazard	Examination of hazard
Biological	Micro-organisms in environment	• Routine environmental monitoring indicates clean rooms are in control. No adverse trends reported.. • Disinfection of clean rooms is performed daily in a consistent and predictable way, which achieves a low bioburden.
	Operators contaminating material (micro-organisms on operating gowns or gloves)	• Operators trained in Aseptic Technique. Aseptic Technique was observed during the simulation and found to be satisfactory. Operators do not directly touch the stoppers. • It is considered that operators can carry out this procedure consistently. • The number of operators involved (two) is suitable for the ergonomics of the operation (see below) but not excessive as to pose a contamination risk. The space around the stoppers is not compromised. It does not compromise HVAC control.

Key Hazard	Source of hazard	Examination of hazard
	Autoclave cycle (presence of bacterial spores)	• Validated physically (thermometrics) and biologically (Biological Indicators) and shown to be repeatable and reproducible.
Physical	Non-viable particles in environment	• Clean rooms are classified six monthly and operate within EU GMP dynamic state conditions. Room pressures are checked daily and there are no recent failures on record. • Maintenance programme is in place to ensure that clean rooms are functioning correctly by testing all HVAC operations.
	Integrity of packaging	• Stoppers move from the autoclave into a sterilised stainless steel container (called a 'hopper'). • The container, once filled, has a stainless lid which is then clamped tight. • Stoppers are not exposed during transit. • Ventilation filters are in place (top and bottom of the hopper)
	HVAC	• Currently functioning correctly. Monitoring systems are in place to detect problems.

Key Hazard	Source of hazard	Examination of hazard
	Integrity of ventilation filters	• Ventilation filters are tested post-use in Sterile Supplies. If filters were to fail, the batch would be put on hold. • Filter is 0.22□m. This will stop ingress of most micro-organisms. The surrounding air is of a low microbial bioburden.
Equipment	Autoclave cycle	• Validated physically (thermometrics) and biologically (Biological Indicators) and shown to be repeatable and reproducible.
	Autoclave cycle failure	• Autoclave cycle parameters checked before opening. Autoclave is alarmed.
	Trolley	• Trolley is manufactured of a material that can be easily cleaned. Composition of material is satisfactory.
Transport	Stopper hopper placed on trolley in Sterile Supplies. The filled hopper is wheeled through a series of clean rooms	• The ergonomics of this operation were assessed. Provided that two members of staff are involved, the transport operation does not expose the stoppers to contamination risk.
	Time	• The time taken (five minutes) is not considered excessive as pose an additional contamination risk
Chemical	None	• There are no known chemical hazards associated with this aseptic transfer

Process flow

The third step is to outline the process flow.

Diagram: Process flow with CCPs (critical control points) marked. Process flow identifies hazards.

Critical Control Points	Process flow
Gowns recently changed	Staff in Suite

Clean room environmental parameters functioning satisfactorily. Tested six monthly as part of HVAC qualification. Room pressures checked daily.	Enter P215
Clean rooms disinfected daily.	

Finger plates will be taken	Staff spray hands with 70% IPA

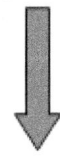

Autoclave cycle validated	Autoclave operation
Using thermometric Probes and Biological Indicators shown to be reproducible and repeatable.	

Under UDAF protection

Autoclave doors are opened

Transfer from drum.

Stoppers are transferred from
Huber Drum into sterilised
stopper hopper.

Hopper (hopper is obtained from autoclaves.

Hopper lid is clamped after stopper transfer.

Hopper has ventilation filters

Minimal manipulations performed

A sterile ventilation filter
is located at top and
bottom of vessel

Vent filter is integrity
Tested.

Trolley is sterilised
autoclaved).

Hopper is fixed onto a trolley (this is

Use of a trolley minimises
Operator manipulations
and increase the speed of
the operation.

Rooms are routinely
corridor to another
environmentally
monitored to Criticality
Factors.

Transit from one autoclave room to

Transit from one corridor to another

No major obstacles
Were observed which
could obstruct the
process flow. Doors can
be opened easily

Positive pressure
airlock differential in place
for airlock

Transit from second corridor to

Transit from airlock to filling room

Stoppers in hopper are attached to Hopper lift,
bottom wrappings are removed. Lift raises hopper to

location under filling machine. Operation from this point onwards is standard.

Environmental monitoring

The fourth step involves assessing the results of dynamic state viable microbiological environmental monitoring and particle counts should be reviewed. If data has been satisfactory and there are no indications of adverse trends, then the proposal can be advanced. If there is a level of risk this must be tackled first (using appropriate corrective and preventative action). This is because undertaking an unusual event in a high risk situation will compound the problem.

Risk Assessment

Perform a simulation

Before undertaking the activity, a simulation should be performed so that any previously unforeseen problems can be noted and further preventative measures taken. If any variability is expected the simulation should be repeated.

The simulation should be timed and the number of staff required noted. The focus should remain on the ease of transit. In this case study the duration of the activity is five minutes and the number of staff involved is two.

Conclusion

From three possible options: high risk, medium risk or low risk, the conclusion is low risk.

This is based on the successful identification of the hazards and control points. The strongest evidence for this came from:

- The process being performed by trained clean room personnel operating in an aseptic manner.
- The autoclaved stoppers being contained within a sealed box. The seal remains intact throughout the transfer.

- The outlet at the bottom of the box has an integrity tested vent filter.
- An additional vent filter is placed on the top of the box.
- The stoppers are wheeled on a trolley, which minimizes human intervention.
- The microbiological environmental monitoring for the rooms is satisfactory.

In developing any risk assessment it is important to note the impact of change and to review any risk assessments on a regular basis.

This case study may or may not be typical and it may or may not be wise for an organization to undertake such an activity. The purpose of this article was to demonstrate an example of the risk assessment approach using a relatively simple case so that the wider application can be appreciated.

References and Further Reading

BS EN ISO 14698 – 1:2003: 'Cleanrooms and associated controlled environments – Biocontamination control – Part 1: General principles and methods'

Friedman, R. L. and Mahoney, S.C. (2003): 'Pharmaceutical cGMPs for the 21st Century: A Risk-Based Approach' in Friedman, R. L. and Mahoney, S.C (eds): 'Risk Factors in Aseptic Processing', Food and Drug Administration, Centre for Drug Evaluation and Research, at: http://www.americanpharmaceuticalreview.com/past_articles/1_AP R_Spring_2003/Friedman_article.htm)

Gavin, A., and L. M. Weddig (eds.). (1995): 'Canned Foods: Principles of Thermal Process Control, Acidification and Container Closure Evaluation', Food Processors Institute,Washington, D.C.

Munro, M. J., Millar, B. W. and Radley. A. S. (2003): 'A Risk Assessment of the Preparation of Parenteral Medicines in Clinical Areas', Hospital Pharmacist, Vol. 10, pp303-305

Notermans, S., Barendsz , A. W. and Rombouts, F. (2002): 'The evolution of microbiological risk assessment in food production', Foundation Food Micro & Innovation, Netherlands

Sandle, T. (2003): 'The use of a risk assessment in the pharmaceutical industry – the application of FMEA to a sterility testing isolator: a case study', European Journal of Parenteral and Pharmaceutical Sciences, 2003; 8(2): 43-49

Sandle, T. (2004): 'General Considerations for the Risk Assessment of Isolators used for Aseptic Processes', Pharmaceutical Manufacturing and Packaging Sourcer, Samedan Ltd, Winter 2004, pp43-47

Whyte, W. and Eaton, T. (2004): 'Microbiological contamination models for use in risk assessment during pharmaceutical production', European Journal of Parenteral and Pharmaceutical Sciences, Vol. 9, No.1, pp11-15

Whyte, W. and Eaton, T. (2004): 'Microbiological risk assessment in pharmaceutical cleanrooms', European Journal of Parenteral and Pharmaceutical Sciences, Vol. 9, No.1, pp16-23

Whyte, W. and Eaton, T. (2004): 'Assessing microbial risk to patients from aseptically manufactured pharmaceuticals', European Journal of Parenteral and Pharmaceutical Sciences, Vol. 9, No.3, pp71-79

6 CASE STUDY – HAZARD ANALYSIS CRITICAL CONTROL POINTS: A NON-STERILE PRODUCTION CASE STUDY (HACCP)

Summary

A full Hazard Analysis Critical Control Point (HACCP) risk assessment has been performed for a non-sterile manufacturing. The HACCP study has been performed to assess the risk from microbiological sources of contamination only.

This report covers the QC Incoming Goods testing.

Scope of Report

In Scope

• QC Goods In Testing, Release and Rejection

Out of Scope

• Liquids Dispensary
• Liquids Manufacturing
• Liquids Packaging
• Warehouse Goods In
• Warehouse Picking
• Warehouse Finished Product
• Warehouse Returns and Quarantine

HACCP Process.

A HACCP process involves:

• A systematic assessment of all process stages.
• Identification of steps critical to product safety.
• Allows concentration of resource on steps critical to product

safety.

HACCP is traditionally presented as 7 linked steps, listed below:

1. Define product and process
    a. The first stage of the risk assessment involves mapping the process for a particular area. Each part of the manufacturing process, raw material and component was mapped on a process flow.
    b. For the process map the entire operation was reviewed with the site experts, in this case the warehouse and QC operators.
2. Identify hazards and control measures
    a. This risk assessment has focused on potential sources of microbiological contamination.
3. Determine Critical Control Points (CCPs).
    a. For each source of microbiological contamination identified the local controls for prevention of that contamination are identified using the simple process flow found below.
4. Establish critical limits for each CCP
5. Establish a monitoring system for each CCP
6. Implement corrective action plan to re-establish control
7. Establish verification procedure to demonstrate compliance

This is summarized in the following schematic:

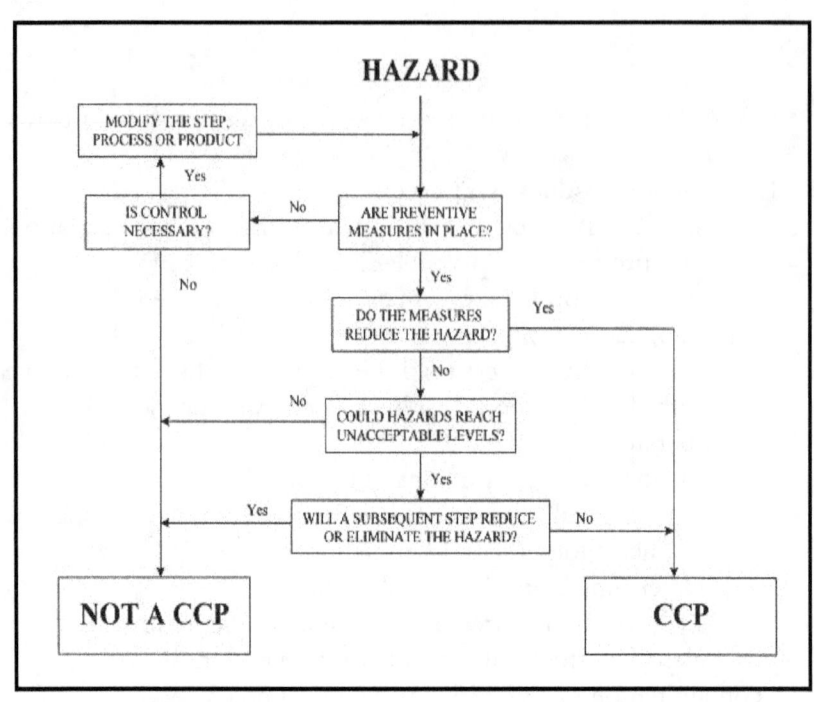

A Critical Control Point (CCP) is defined as a step which, if controlled, will eliminate or reduce a hazard to an acceptable level.

An example is a case study for a QC Goods In Testing, Release and Rejection process. The object of the HACCP is to identify critical control points (CCP). When CCPs occur an action is required: either a risk assessment and an associated attempt at risk mitigation, or some form of monitoring.

The areas to consider include:

Process stage	Hazard
Sample Receipt	External contamination enters sampled material
	Incorrect sampling of raw material leading to false negative result
	Incorrect sampling of raw material leading to false negative result
	Incorrect sampling of raw material leading to false negative result
	Microbial contamination of material to be sampled

Process stage	Hazard
	Use of wet pallet
	No plastic pallet available
Use of Plastic Pallets	Contaminated plastic pallet used
	Damaged wooden pallet (part) transferred into sample booth

Process stage	Hazard
Transfer materials to airlock	Contaminated outer packaging
	Contaminated carrier
	Airlock door failure
	Contaminated outer packaging

Process stage	Hazard
Transfer materials to airlock	Contaminated outer packaging
	Contaminated carrier
	Airlock door failure
	Contaminated outer packaging
	AHU Failure

Process stage	Hazard
Sample Booth Air Handling Unit (AHU)	Booth airflow system not turned on
	Booth airflow too low
	Booth airflow too high
	Airflow patterns not established before sampling begins
	Booth airflow system not turned on during sampling
	Damaged HEPA filter in Booth

Process stage	Hazard
Operator activities	Untrained operator performing sampling
	Operator not gowned correctly
	Gloves not sanitized on entry
	Gloves not worn
	Unauthorized access into booth
	Untrained operator performing sampling

Process stage	Hazard
Booth changing room	Airlock interlock not operating correctly
	changing room dirty

Process stage	Hazard
Sampling	Sample Booth dirty
	Sample equipment dirty
	Sample not closed properly
	Opening of contaminated sample
	Use of contaminated sample equipment
	Sample Booth dirty
	Use of contaminated sealing tape
	QC sample taken before micro sample
	Operator paperwork contaminated
	No sterile sample pots available

Process stage	Hazard
Sample Booth Cleaning	Poor cleaning performed
	No cleaning performed
	Use of contaminated vacuum cleaner
	Use of contaminated cleaning equipment
	Use of uncontrolled water
	Poor cleaning performed
	No cleaning performed
	Use of contaminated vacuum cleaner
	Use of contaminated cleaning equipment
	Use of uncontrolled water

Once the first stage of the HACCP has been completed, the critical control points (CCPs) require examination.

HACCP Critical Control Points

Some examples in relation to materials.

Step	Control Points
Transfer Materials to Airlock	Environmental monitoring program to be put in place for the booth
Transfer Materials to Airlock	Dedicated carrier required. Larger sample booth
Sample Booth AHU #1	Time delay required between turning the booth on and using it
Sample Booth AHU #2	Documented check that the in use HEPAs are within test date.
Operator Activities	Introduction of alcohol glove sanitization
Booth Changing Room	Document routine cleaning
Sampling #1	Dedicate knife to sample booth
Sampling #2	Always sample microbiological samples first
Sampling # 3	Have a dedicated area for paperwork
Sample Booth Cleaning #1	Routine cleaning required for vacuum cleaner - routine change of HEPA filters
Sample Booth Cleaning #2	Introduce increased control of cleaning equipment
Sample Booth Cleaning #3	Review cleaning standards for booth and sample water point for microbial bioburden.

Once control measures have been identified and analyzed, an implementation plan to address corrective measures is required. Once a HACCP has been completed, it should be subject to regular review to ensure that control measures continue to be achieved.

7 CASE STUDY - RISK ASSESSMENT OF A STERILITY TESTING ISOLATOR (FMEA)

Abstract

Risk assessment is an increasingly used analytical tool within the pharmaceutical industry. One such method is FMEA (Failure Mode and Effects Analysis). This paper uses an FMEA approach to examine a sterility testing Isolator by studying it for severity of risk, occurrence of risk and method of detection of the risk and thereby shows the usefulness of FMEA for this study and for its wider applications to other areas of testing and manufacturing.

Introduction

The use of risk assessment in the pharmaceutical industry is both an increasingly used tool and an expectation of regulatory authorities. The use of risk assessments is an important process to allow manufacturers to justify current practices, to explore weaknesses and to construct rationales. Risk assessment tools also provide a means for the validation of processes (such as the approach referred to in the FDA Code of Federal Regulations, CFR 21, Part 820).

There are several different approaches for risk assessment including HACCP (Hazard Analysis Critical Control Points), which has its origins in the food industry; Fault Tree Analysis (FTA) ; modeling software (such as the Monte Carol model); and FMEA (Failure Mode and Effects Analysis), which originated in the engineering sector. Of these, HACCP and FMEA focus upon answering the question: 'what will happen if a failure occurs?', whereas FTA asks the questions 'what caused the failure to happen?'

Sometimes these approaches form part of a company's total quality system, sometimes they exist as standalone techniques. These techniques are among others described in ICH Q9 'Quality Risk

Management'. This document provides the structure for various risk analysis approaches which are centered on the question: 'what can go wrong?' Similar tools are also described in the FDA concept paper 'Pharmaceutical GMPs for the Twenty-First Century: A Risk Based Approach'.

Formal risk approaches normally share four basic concepts, which are listed below:

- **Risk assessment;**
- **Risk control;**
- **Risk review;**
- **Risk communication.**

Before commencing a risk assessment it is important to define the size and the scope of the assessment (remaining focused on what is to be achieved); to select the appropriate team (often an interdisciplinary team is best); selecting and reviewing the appropriate risk management tool; deciding upon any numerical scale to be used and prioritizing the different problems to be addressed. Most approaches begin by constructing a process map.

These steps can be broke down thus:

- Gathering data through an audit and analysis;
- Constructing diagrams of work flows;
- Pin-pointing areas of greatest risk;
- Examining potential sources of contamination;
- Deciding on the most appropriate sample methods;
- Helping to establish alert and action levels;
- Taking into account changes to the work process / seasonal activities;
- Using some type of scoring system so that the risk can be ranked and the level of risk determined.

In doing so the following questions should be asked:

- What is the function of the equipment? How are its performance requirements?

- How can it fail to fulfil these functions?
- What can cause each failure?
- What happens when each failure occurs?
- How much does each failure matter? What are its consequences?
- What can be done to predict or prevent each failure?
- What should be done if a suitable proactive task cannot be found?

The use of risk assessment techniques can also be used to determine the type, locations and frequencies of monitoring required to give the Isolator user confidence that the system is operating satisfactorily (Sandle, T: 2003).

When examining different risks it is often appropriate to determine the level of monitoring required. However, it must be stressed that monitoring is a means of detecting risks and should be used primarily where a risk cannot be completely removed or there is a possibility of the mechanism to avoid that risk breaking down. It should never be used solely in place of a detailed risk assessment. However, dynamic state microbiological environment monitoring can never be avoided either: no risk assessment approach can eliminate the need to monitor entirely and there will be a need to consider the following:

1. Where monitoring takes place?
2. How often monitoring is performed?
3. Appropriate corrective and preventative actions for action level excursions.

The types and locations for monitoring must have relevance to the process, that is, the data produced must be meaningful.

The requirement to perform microbiological monitoring is derived from the following standards and guidelines:

- EU GMP Guide (the 'Orange Guide')
- USP <1116>
- FDA Code of Federal Regulations 21 CFR 211
- BS EN ISO 14698 – 1:2003

Environmental monitoring typically consists of the following sample types:

1. Passive air-sampling: settle plates
2. Active air-sampling: volumetric air-sampler
3. Surface samples: contact (RODAC) plates
4. Surface samples: swabs
5. Finger plates
6. Plates of sleeves / gowns
7. Particle counting

The locations for monitoring should be indicated on a sampling map so that sampling is consistent and reproducible.

This chapter examines FMEA as a tool and uses the example of a barrier Isolator system used for sterility testing by applying FMEA as a tool for a risk assessment of an Isolator system. FMEA has been selected as the risk assessment approach partly based on the draft PIC/S guideline on inspecting Isolator systems:

"There should be evidence of documented Risk and Failure Mode analysis of the Isolator and its process systems"

Failure Mode and Effects Analysis

Failure mode and effects analysis or failure modes and effects criticality analysis, as it is sometimes alternatively known, (FMEA / FMECA) is an analytical tool which originated in the engineering industry (for example, refer to the public domain information presented by the engineering multinational company ABB . FMEA is now being applied to other areas as a problem solving tool and this paper illustrates this by examining an Isolator system. There are many approaches the performing of an FMEA analysis. This paper adopts one possible approach based on a general introduction to FMEA detailed by the non-commercial FMEA Information Centre (The FMEA centre at: http://www.fmeainfocentre.com/introductions.htm).

FMEA is a highly structured approach and can be undertaken through the following steps:

1. Setting the scope;
2. Defining the problem;
3. Setting scales for factors of severity, occurrence and detection (see below);
4. Process mapping;
5. Defining failure modes;
6. Listing the potential effects of each failure mode;
7. Assigning severity ratings to each process step;
8. Listing potential causes of each failure mode;
9. Assigning and occurrence rating for each failure mode;
10. Examining current controls;
11. Examining mechanisms for detection;
12. Calculating the risk;
13. Examining outcomes and proposing actions to minimize risks.

Where the number of risk is very high, the ICH Q9 guidelines proposes the use of a risk filter.

Sterility Testing Isolator: The case study

The definition of an Isolator, taken in this article, is a device:

• Provided with microbially retentive filtered air (and which does not exchange any other air with the surrounding environment)
• Has a decontamination cycle (for the Isolator itself and for material entering)
• Has a means for material transfer and / or connection to another Isolator
• No human part directly enters the Isolator

(Reference: PDA Technical Report No.34).

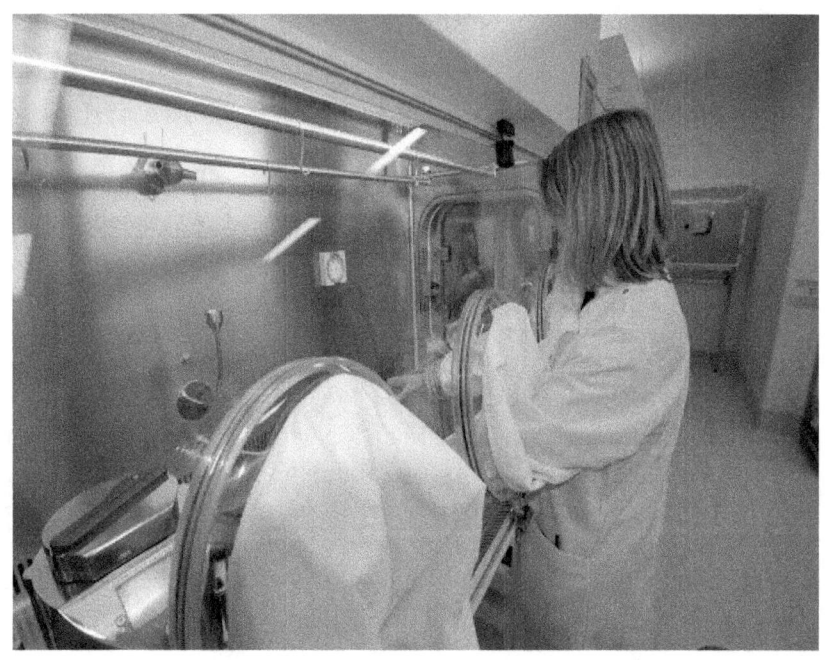

All Isolators are at risk from contamination. Although Isolators are superior in many ways to clean rooms, the approach of regulators, such as the FDA, is:

"Barrier Isolators cannot prevent contamination caused by GMP deficiencies such as poor aseptic procedures and inadequate training of...operators"

(The Gold Sheet, Vol. 32, No.10, October 1998)

The operation of Isolators associated with aseptic processing have a very good history in terms of minimizing contamination. These Isolators are Type 1 positive pressure Isolators, designed to meet EU GMP Grade A at rest, and tend to be constructed using a barrier of either flexible film (such as PVC) or a rigid wall (such as stainless steel grade 304, 316L) (Lee & Midcalf, 1994).

Description of the system

The sterility testing Isolator system at examined is from a pharmaceutical manufacturer based in the south-east of England. The Isolator primarily consists of one half-suit Isolator, two transfer Isolators and a steriliser unit. The Isolators were manufactured by La Calhene and are positive pressure, flexible film Isolators with stainless steel frames and wood bases designed for aseptic processes (in this case: sterility testing). Air is supplied into the Isolators using microbiologically retentive filtration using HEPA filters and materially is transferred into and out of the main Isolator using transfer Isolators connected using Rapid Transfer Ports (RPT). The system for sanitization is peracetic acid (3.5% w/v). The inside of each Isolator, in operation, is classed as Grade A (as defined by EU GMP 2002).

The Isolators are approximately fifteen years old, although they have undergone several structural changes including a regular six monthly change of the main Isolator flexible film canopy. The use of the wood basis and the peracetic acid sanitization cycle are historical and based on the technology of the time. It is likely that if a system was being purchased today a system with a stainless steel bases and an arguably faster vapor phase hydrogen peroxide (VHP) sanitization system would be chosen.

The main risks which different Isolators (those used for both sterility testing and for aseptic filling) are susceptible include:

• Leaks;
• Gloves / operator manipulations;
• Filters;
• Other airborne contamination;
• Transfer of material into and out of the Isolator;
• The Isolator room;
• Decontamination cycle;
• Cleaning / environmental monitoring issues.

Some of these risks are examined through the FMEA below.

Application

The Isolator system is used for the sole purpose of performing final product sterility testing on a range of plasma derived parenteral products according to Ph. Eur. 2.6.1. The methods used are membrane filtration and direct inoculation. A variety of environmental monitoring methods are performed during and after testing: air-samples (passive settle plates and an active volumetric air-sample); finger plates; contact plates and swabs. A spray bottle of a sporicidal disinfectant remains in the Isolator for spillages and for a post-test clean down. The system has a good history of microbiological monitoring results with isolated colonies being detected infrequently. Monthly monitoring is performed in the Isolator room. This is not a regulatory requirement and the room is not classified. However such monitoring is performed in order to detect any adverse trends and to allow any required action to be taken.

In addition to the viable microbiological monitoring, A number of daily, weekly and six-monthly physical tests are performed on the Isolator system using pressure charts; cleaning and formal classification as a Grade A clean zone (to ISO 14644-1).

The FMEA study on the sterility testing Isolator system

The pharmaceutical company using the Isolator undertook to perform a retrospective risk assessment on its sterility testing Isolator system using FMEA as the most appropriate tool.

Designing the FMEA scheme

FMEA schemes vary in their approach, scoring and categorization. All approaches share in common a numerical approach. The approach adopted at the organization was to assign a score (from 1 to 5) to each of the following categories:

i) Severity
    ii) Occurrence
    iii) Detection

    Where:

    i) Severity is the consequence of a failure, should it occur;
    ii) Occurrence is the likelihood of the failure happening
        (based on past experience);
    iii) Detection is based on the monitoring systems in place
        and on how likely a failure can be detected.
    Sometimes, a good detection system is described
as one that can detect a failure before it occurs.

By asking a series of questions of each main part of the Isolator system. Such questions included:

i) What is the function of the equipment? How are its performance requirements?
ii) How can it fail to fulfill these functions?
iii) What can cause each failure?
iv) What happens when each failure occurs?
v) How much does each failure matter? What are its consequences?

vi) What can be done to predict or prevent each failure?

vii) What should be done if a suitable proactive task cannot be found?

The scoring system adopted was: a scale from 1 to 5. It followed that the likelihood of high severity would be rated 5; high occurrence rated 5; but a good detection system would be rated 1. The scoring system was based on the table below:

Severity	5	Specification limits exceeded. Probable rejection of test or shutdown of system.
	3	Observed trend takes place, but no critical excursions. Requires investigation.
	1	No excursion has taken place. No upward trends and no investigation is required.
Occurrence	5	Expected to occur ☐50% time.
	3	Expected to occur ≥10 - ≤50% time.
	1	Expected to occur ≤10%.
Detection	5	No way to detect the failure mode.
	3	Can be partially detected but detection could be improved.
	1	Good detection systems in place.

Some organizations prefer to use a 10 point scale.

Using these criteria a final FMEA score is produced (sometimes called a Risk Priority Number):

$$\frac{x}{125}$$

The total of 125 is derived from: severity score x occurrence score x detect score, or:

$5 \times 5 \times 5 = 125$

Depending upon the score produced it can be decided whether further action is needed. The is no published guidance on what the score that dictates some form of action should be. In this study the company adopted a score of 27 as the cut-off value where action was required. This was based on 27 being the score derived when the mid-score is applied to all three categories (i.e. the numerical value '3' from severity (3) x occurrence (3) x detect (3)) and the supposition that if the mid-rating (or a higher number) was scored for all three categories then as a minimum the system should be examined in greater detail.

The FMEA exercise

To conduct the exercise used the defined scheme on the Isolator system, the Isolator set-up was broken down into a number of critical areas and each area was subsequently assessed. Each main step is detailed and examined below.

Examination 1: The Isolator room

Description of critical area: The Isolator is situated in an unclassified room. There is no requirement to place a sterility testing Isolator in a classified room.

FMEA schematic:

Process step	Failure Mode	Significance of failure	Severity of consequence (**score**)
Loading Isolators pre-sanitisation / performing sterility testing	That contamination from the room could enter transfer or main Isolators	Reduced efficiency of transfer Isolator sanitisation / contamination inside main Isolator	3

Measures to detect failure	Occurrence (**score**)	Detection systems	Detection (**score**)
Would be shown from reduced evaporation rate for Isolator sanitisation / poor environmental monitoring results in main Isolator / potential sterility test failures / sanitisation cycle has been validated using BIs of 10^6 spores	1	Isolator room is monitored monthly for viable micro-organisms and papers / staff wear over-shoes on entry / Dycem mat in place / entry to room has controlled access / environmental monitoring performed inside main Isolator / Isolators are at positive pressure to the room and air is HEPA filtered	1

FMEA score: $3 \times 1 \times 1 =$ 3

Analysis: There is no problem considered from the room environment. Entry to the room is controlled; the sanitization cycle has been challenged with a level of micro-organisms far greater than would ever be found in the environment (spores of *Geobacillus stearothermophilus*); all items entering the Isolator are sanitised (using a chlorine dioxide based sporicidal disinfectant) and the Isolator itself is an effective positive pressure barrier to the outside (at >15 Pascal).

As detailed earlier, environmental monitoring is performed inside the Isolator during testing. This monitoring, which has an action level of 1 cfu, is designed to detect any potential contamination inside the Isolator environment.

Examination 2: Potential of sanitisation cycle failure

FMEA schematic:

Process step	Failure Mode	Significance of failure	Severity of consequence (**score**)
Performing sanitisation cycles on transfer or main Isolator	An Isolator is not correctly sanitised	Contaminated items enter main Isolator or main Isolator itself is contaminated	4

Measures to detect failure	Occurrence (**score**)	Detection systems	Detection (**score**)
Evaporation rate / pre- and post-lot testing of acid / sanitisation cycles developed using BIs	1	Steriliser parameters checked after sanitisation and before use / acid potency checked for each lot / post sanitisation environmental monitoring performed for main Isolator	1

FMEA score: 4 x 1 x 1 =   4

Analysis: The severity of an ineffective sanitsation cycle is a potential sterility test failure. However the steriliser parameters are checked for every transfer and main Isolator cycle and post-sanitisation environmental monitoring is performed on the main Isolator. This has a long history of producing no growth of viable micro-organisms.

The Isolators are loaded with a set amount of equipment and consumables. This is described in authorized procedures and the maximum load has been determined through BI studies.

One potential area of weakness for the sanitization of the main Isolator are valves for the removal of waste during the membrane filtration sterility test. These are autoclaved prior to each sanitization and during the first hour of the cycle they are opened both inside and outside to allow the sanitization agent to penetrate. A further preventative measure is taken post-sterility testing where the valve which has been used is rinsed through with disinfectant.

Examination 3: frequency of Isolator sanitizations

FMEA schematic:

Process step	Failure Mode	Significance of failure	Severity of consequence (**score**)
Performing sanitisations on transfer (each batch) and main Isolator (three monthly)	Isolators are not sanitised frequently enough and allow contaminatio n build up	Environment inside Isolator becomes contaminated thereby increasing likelihood of sterility test failure	4

Measures to detect failure	Occurrence (**score**)	Detection systems	Detection (**score**)
Environmental monitoring inside main Isolator / physical checks	2	Analysis of environmental monitoring / physical checks performed daily & weekly & six-monthly service and calibration	1

FMEA score: 4 x 2 x 1 =   8

Analysis: Each transfer Isolator is sanitised each run using a validated cycle and the Isolators are cleaned monthly to remove grease, debris and so on. The sanitisation physical parameters are checked each run (evaporation rate and pressure chart recorder).

The main Isolator is sanitised every three months (this has been set by monitoring trends in biocontamination over time). Environmental monitoring is performed during each sterility test and examined monthly for trends. If micro-organisms are detected from sterility test environmental monitoring the area of detection is sanitised. Should contamination re-occur the area is again disinfected. If contamination occurs again, or is of a high level, the Isolator is re-sanitised.

It has been shown that if contamination occurs at a location in the main Isolator it does not re-occur when repeat monitoring is performed. It is reasoned that this is because the level of post-test disinfection is sufficient; that the air-changes in the Isolator are such that most contamination will be removed every hour and that the environment within the Isolator is such that it is generally inhospitable to the survival and reproduction of vegetative micro-organisms.

Furthermore, the main Isolator is continuously monitored to show that it remains at positive pressure to the outside and every six-months a range of physical tests are performed: pressure decay, HEPA filter integrity and paper classification.

Examination 4: compromise of Isolator integrity

FMEA schematic:

Process step	Failure Mode	Significance of failure	Severity of consequence (**score**)
Use of transfer Isolators / performing sterility tests	Isolator port fall off exposing inside of Isolator to external environment / large tear to fabric of Isolator / cut to Isolator glove	Potential contamination inside Isolator or pressure drop	4

Measures to detect failure	Occurrence (**score**)	Detection systems	Detection (**score**)

Measures to detect failure	Occurrence (**score**)	Detection systems	Detection (**score**)
Environmental monitoring / pressure monitoring / visual observation	2	Environmental monitoring will show contamination / pressure charts show drop / technicians can respond to ports falling off or cuts to gloves / air changes aid removal of contamination / post-test disinfection performed in main Isolator	2

FMEA score: 4 x 2 x 2 = 16

Analysis: It is generally recognised that, despite being a barrier system at positive pressure, any Isolator will have some leakage. The issue of this FMEA is on wider tears or breaches to the integrity of the Isolator system. An occurrence of 2 has been given because there have been a small number of isolated incidents where a transfer port has fallen off prior to connection to the main Isolator or that a tear to a glove has occurred.

Should an Isolator develop a small tear it will remain at positive pressure. This has been shown by small holes in the Isolator being detected during some six-monthly calibrations where small holes have been detected. Despite this there have been no sterility test failures or adverse environmental monitoring failures. It is therefore considered that small holes to the fabric of the Isolator, while undesirable, present a very low risk.

Should a large tear be detected the Isolator is decommissioned, the canopy replaced and the Isolator is re-sanitised. For the main Isolator a spare canopy is normally held at The company because a each canopy is made to order and this will take approximately six weeks to manufacture.

As a preventative measure the canopy of the main Isolator is replaced once per year to avoid any small holes which may develop during the course of the year becoming a problem. The transfer canopies are reviewed once per year and replaced when considered to be wearing. Because the organisation has two transfer Isolators it does not consider the cost of having a spare transfer canopy to be justifiable because this will not halt sterility testing while the canopy is replaced.

If a glove was torn during a sterility test it would be replaced following a defined procedure. The testing technician wears a pair of sterile gloves underneath the Isolator gloves. This is considered to give some measure of protection while the glove is being changed. In such an event a Microbiology Investigation report would be generated and the results of the environmental monitoring examined. If the monitoring was satisfactory this would be noted, if the monitoring was unsatisfactory consideration would be made to the most appropriate action: disinfection inside the Isolator or decommissioning and re-sanitisation. Unless there was a sterility test failure no further action would be taken.

If a transfer port fell off from a transfer Isolator or from the main Isolator the Isolator would be decommissioned and re-sanitised.

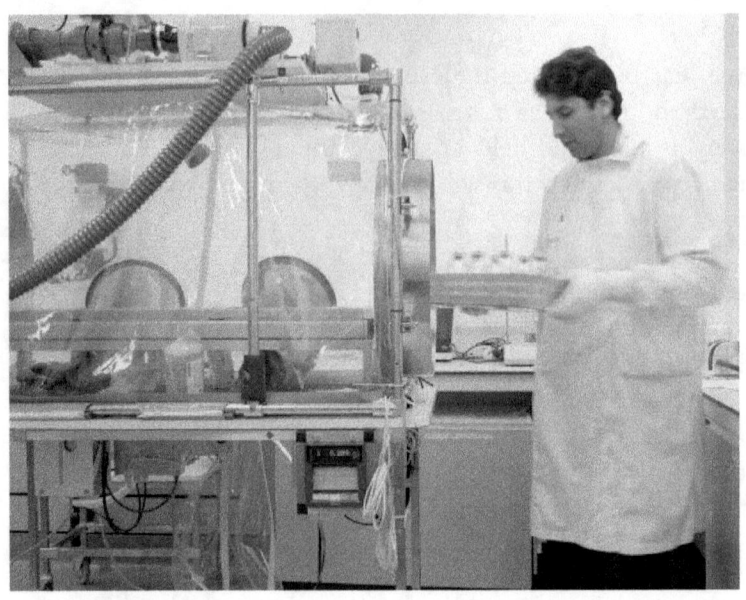

Examination 5: connection of transfer Isolator to main Isolator and transfer-in / out of material

The transfer of material in and out of the Isolator is, arguably, the biggest risk:

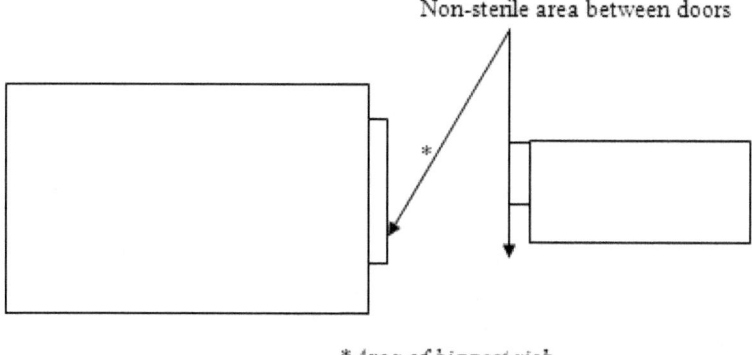

Non-sterile area between doors

Area of biggest risk

Diagram 1: illustration of a connection of two Isolators, with the 'ring of concern'.

FMEA schematic:

Process step	Failure Mode	Significance of failure	Severity of consequence (**score**)
Connection of transfer Isolator to main Isolator and moving material in and out	Contamination on outside of both Isolators may enter the main Isolator / failure to maintain positive pressure	Contamination enters the Isolator or positive pressure is not maintained	4

Measures to detect failure	Occurrence (**score**)	Detection systems	Detection (**score**)
Environmental monitoring / pressure monitoring	1	DPTE seal system / use of disinfectant for connection	1

FMEA score: 4 x 1 x 1 =   4

Analysis: It is generally accepted that the two theoretically weakest points on an Isolator system are the transfer of materials in and out of an Isolator and the Isolator gloves. This FMEA has examined the transfer of materials in and out.

Transfer of material into a transfer is of low risk because the items are all wiped with a disinfectant by staff wearing sterile gloves and because of the demonstrated effectiveness of the sanitisation cycle.

Transfer of material into and out of the main Isolator from a transfer Isolator is done via DPTE ports which ensure that surfaces exposed to the outside environment are sealed in a way that they do not come into contact with the inside of the Isolator. Prior to connection the transfer ports of both Isolators are wiped with disinfectant and during connection both Isolators remain at positive pressure. During testing both Isolators remain connected (although the port is replaced in order to keep the pressure at a higher level. The pressure when the ports are opened remains positive but at a lower level. Therefore in order to reduce the risk of pressure drops the link between the two Isolators only remains open for a short period of time for transfer in and transfer out).

Experience has shown that positive pressure is always maintained (from chart recorders) and environmental monitoring of the port and a location close by in the main Isolator has a good history.

All tested material and waste is placed back into the transfer Isolator after testing (with measures to ensure that any sharp objects are rendered safe to avoid any breach to the Isolator's plastic film). A waste container used to be affixed to the Isolator but this was removed following difficulties in demonstrating an effective sanitisation.

Examination 6: incomplete transfer Isolator sanitisation

FMEA schematic:

Process step	Failure Mode	Significance of failure	Severity of consequence (**score**)
Performing Isolator sanitisation	Sanitisation not completed or items not moved around half-way through cycle	Incomplete sanitisation	3

Measures to detect failure	Occurrence (**score**)	Detection systems	Detection (**score**)
Cycle times and moving of items mid-cycle reorded; evaportation rate recorded; amount of sterilant added recorded	2	All parameters recorded and checked by another technician	1

FMEA score: 4 x 2 x 1 =   8

Analysis: If a sanitisation cycle was aborted it would be repeated. The time of the start and end of the cycle is recorded and physical checks are recorded (positive pressure and the evaporation rate of the sterilant) so that an incomplete cycle could be seen.

The sanitisation cycle for the transfer Isolators has shown a weakness with the sanitisation cycle in that it is only truly effective at destroying viable and spore bearing micro-organisms when the sterilant (peracetic acid) comes into direct contact with the surface. In order to ensure this happens all items in a transfer Isolator are moved, in such a way that previous unexposed surfaces are exposed, half way through sanitsation. The time is recorded by the technician. If this did not occur the Isolator would be re-sanitised.

Examination 7: failure of a daily, weekly or six-monthly physical parameter - HEPA filters / pressure leaks to canopy

FMEA schematic:

Process step	Failure Mode	Significance of failure	Severity of consequence (**score**)
Performing physical checks at defined intervals	HEPA filter failure or pressure leak	Possibility of microbial ingress	4

Measures to detect failure	Occurrence (**score**)	Detection systems	Detection (**score**)
Environmental monitoring performed regularly / pressure checked each working day / HEPA filters and pressure decay performed six-monthly	1	The frequent environmental monitoring and pressure checks provides high level assurance / HEPA filters; pressure; inspection of filter housing; and air-changes checked as part of six-monthly preventative maintenance programme	1

FMEA score: $4 \times 1 \times 1 = 4$

Analysis: The six-monthly preventative maintenance programme for the Isolator has not shown a significant leak for the main Isolator. HEPA filters have been replaced as part of a preventative measure when paper counts have shown an occasional elevation.

It is considered that the six-monthly frequency of the physical tests is satisfactory given the good history of the system and the low incidence of related problems. Frequent pressure monitoring and environmental monitoring provide sufficient alert system for any micro-organisms being present in the Isolator environment.

Examination 8: pressure leaks to gloves

FMEA schematic:

Process step	Failure Mode	Significance of failure	Severity of consequence (**score**)
Use of gloves to transfer material or to perform sterility test (sterile gloves may be worn underneath Isolator gloves)	Contamination from technician into Isolator or weak area of positive pressure to allow contamination in	Contamination present in Isolator / compromise of aseptic technique	4

Measures to detect failure	Occurrence (**score**)	Detection systems	Detection (**score**)
Environmental monitoring (post-use finger plates) / pressure charts	2	Environmental monitoring is performed post-test on gloves / gloves are wiped with disinfectant / gloves are visually examined weekly and changed as	3

Measures to detect failure	Occurrence (**score**)	Detection systems	Detection (**score**)
		appropriate	

FMEA score: 4 x 2 x 3 = 24

Analysis: This FMEA as been given a occurrence of 2 because weekly checks on the gloves do show, on occasions, holes in gloves. A detection of 3 has been given due to reasons outlined below.

The weakest spot on the Isolator is considered to be the glove ports therefore the gloves have been given a separate FMEA. Although these are tested after each test using finger plates and are visually inspected by the testing technician pre-test and weekly, such visual checks are unable to detect pin-pricks leading to slow leakage. Pressure monitoring would show a significant leak from torn gloves but is not subtle enough to detect tiny holes. In order to improve detection the organisation undertook to purchase a glove-leak tester. This reduced the FMEA score by improving the detection rate from 3 to 1.

The probability of contamination is further reduced by the use of aseptic technique by the testing technician at all times. Tests are performed to the same level of aseptic technique that would be provided to performing a sterility test in a clean room. Furthermore all technicians are trained in aseptic technique prior to testing final product for batch release.

In addition technicians wear a pair of sterile gloves underneath the Isolator gloves and procedures are in place for an aseptic change of gloves. Spare gloves are held in the Isolator for this purpose.

Despite the pre-glove leak testing system FMEA rating of 24 - as a possible risk - the good history of environmental monitoring gives assurance that there is little contamination in the Isolator and no adverse trends. Therefore the gloves are a potential weak spot but this has not been observed in practice. A further weakness is associated with the glove change procedure, which could also be explored.

Other leaks associated with the Isolator also pose a risk and could be similarly examined trough FMEA. Such issues can include the following:

Leaks / pressure

Identification of risk	Minimising the risk	Monitoring the risk
• Leaks from gloves, fabric or sleeves. • High risk is a leak exceeding 1.5% of Isolator volume. • Risk increases with larger hole; if the background environment in contaminated or if the positive pressure of the Isolator relative to the room is not high enough.	• The flexible wall of the Isolator needs to be thick enough to avoid puncture • The type and nature of seals and joints should be appropriate. • The impact of any potential leak should be risk assessed in advanced. • Isolator to be maintained at a positive pressure (>10 Pascals)	• Build in pressure testing / leak monitoring as part of the preventative maintenance programme (such as the pressure decay test or tracer gas detection test). This could be checked six monthly or prior to running a decontamination cycle • Have constant pressure monitoring of the Isolator with alarms

A further series of risks could be associated with a general contamination build up. Such an examination would be useful in determining the frequency of sanitization inside an Isolator environment.

Conclusion

This paper has demonstrated that a risk assessment technique – FMEA can be readily applied to a key operation within it: namely a sterility testing Isolator. What is also of interest is that this technique did not originate in the pharmaceutical industry. This indicates how the synergy of different approaches can be achieved.

The FMEA approach can be summarised graphically. This can be useful for tracking trends:

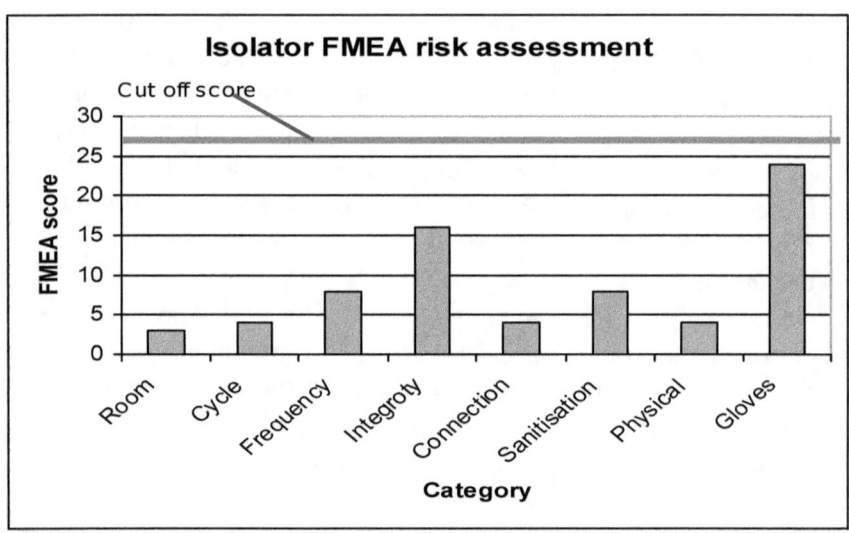

By using FMEA the Isolator system has been examined for potential risks, the severity of those risks, the likelihood of those risks occurring in practice and the means of detection. Such an analysis can prove to a useful tool for the pharmaceutical company in examining its own systems and in preparing a rationale for regulatory bodies.

References and further reading

Ackers, J., Agalloco, J. and Kennedy, K. (1995): 'Experience in the Design and Use of Isolator Systems for Sterility Testing', PDA Journal of Pharmaceutical Science and Technology, Vol. 49, No.3, May-June 1995, pp140-144

Agalloco, J. (1999): 'Barriers, Isolators and Microbial Control', PDA Journal of Pharmaceutical Science and Technology, Vol. 53, No.1, January-February 1999, pp48-53

BS EN ISO 14698 – 1:2003: 'Cleanrooms and associated controlled environments – Biocontamination control – Part 1: General principles and methods'

Code of Federal Regulations, 1998, Title 21, Part 210, Current Good Manufacturing Practice in Manufacturing, Processing, Packaging, or Holding of Drugs – General, 210:3

Code of Federal Regulations, Title 21, Food and Drugs, Part 820, Quality Management Regulations, 820.75

De Abreu, C., Pinto, T. and Oliveira, D. (2004): 'Environmental Monitoring: A Correlation Study Between Viable and Nonviable Particles in Clean Rooms', Journal of Pharmaceutical Science and Technology, Vol. 58, No.1, January-February 2004, pp45-53

Farquarharson, G and Whyte, W. (2000): 'Isolators and Barrier Devices in Pharmaceutical Manufacturing', PDA Journal of Pharmaceutical Science and Technology, Vol. 54, No.1, January-February 2000, pp33-43

FDA: 'Guidelines on Sterile Drug Products Produced by Aseptic Processing', FDA, 1987 (draft produced for review in August 2003)

Kaye, S. (1986): 'Efficiency of Biotest RCS as a Sampler of Airborne Bacteria', Journal of Parenteral Scicence and Technology, Vol. 42, No.5, September-October, pp147-152

Lee, G and Midcalf, B. (eds): 'Isolators for Pharmaceutical Applications', 1994, HMSO

International Conference on Harmonisation. Q9, Quality Risk Management, June 2006.

Ljungqvist, B. and Reinmuller, B. (1996): 'Some observations on Environmental Monitoring of Cleanrooms', European Journal of Parenteral Science, 1996, 1: 9 –13

Mackler, S. (2000): 'Barrier Isolation Technology: Facilities Update', Pharmaceutical Technology, February 2000, pp40-47

Meir, R. and Zingre, H. (2000): 'Qualification of air-sampler systems: MAS-100', Swiss Pharma, 22(2000); pp15 - 21

Neiger, J. (1997): 'Life with the UK pharmaceutical Isolator guidelines: a manufacturer's viewpoint', European Journal of Parenteral Sciences, 1997, Vol.2, No.1, pp13-20

Ohresser, S., Griveau, S. and Schann, C. (2004): 'Validation of Microbial Recovery from Hydrogen Peroxide-Sterilised Air', Journal of Pharmaceutical Science and Technology, Vol. 58, No.2, March-April 2004, pp75-80

PDA Technical Report No. 13 (revised): 'Fundamentals of an Environmental Monitoring Programme', September / October 2001

PIC/S – Recommendations on the Inspection of Isolator Technology (draft), 2001

PhRMA Environmental Monitoring Work Group (1997): 'Microbiological Monitoring of Environmental Conditions for Nonsterile Pharmaceutical Manufacturing', Pharm. Technol., March, pp58-74

Reich, et al. (2003): 'Developing a Viable Microbiological Environmental Monitoring Program for Nonsterile Pharmaceutical

Operations', <u>Pharm. Technol.</u>, March, pp92-100

'Rules and Guidance for Pharmaceutical Manufacturers and Distributors' ('EU GMP Guide'), MHRA, 2002

Sandle, T.: 'Environmental Monitoring in a Sterility Testing Isolator', <u>PharMIG News</u> No.1, March 2000

Sandle, T.: 'Microbiological Culture Media: Designing a Testing Scheme', <u>PharMIG News</u> No.2, August 2000

Sandle, T. (2003): 'The use of a risk assessment in the pharmaceutical industry – the application of FMEA to a sterility testing Isolator: a case study', <u>European Journal of Parenteral and Pharmaceutical Sciences</u>, 2003; 8(2): 43-49

Sandle, T (2003)[2].: 'Selection and use of cleaning and disinfection agents in pharmaceutical manufacturing' in Hodges, N and Hanlon, G. (2003): '<u>Industrial Pharmaceutical Microbiology Standards and Controls</u>', Euromed Communications, England

Sandle, T. (2004): 'General Considerations for the Risk Assessment of Isolators used for Aseptic Processes', <u>Pharmaceutical Manufacturing and Packaging Sourcer</u>, Samedan Ltd, Winter 2004, pp43-47

Sidor, L. and Lewus, P. (2007): 'Using risk analysis in Process Validation', <u>BioParm International,</u> pp50-57

US Food and Drug Administration Pharmaceutical GMPs for the 21st Century – A Risk Based Approach. Final Report, Rockville, MD. September 2004

USPNF#25 <1116>

Wagner, C. (1995): 'Current Challenges to Isolation Technology' in Wagner, C. and Ackers, J. (eds): 'Isolator Technology: Applications in the Pharmaceutical and Biotechnology Industries'. 1995, Interpharm Press

Whipple, A. (1999): 'Practical validation and monitoring of Isolators used for sterility testing', European Journal of Parenteral Sciences, 1999, Vol. 4, No.2, pp49-53

Whyte, W. (2001): 'Cleanroom Technology: Fundamentals of Design, Testing and Operation', Wiley, Chichester

Wilkins, J. (1994): 'The advantages of aseptic filling in a rigid Isolator', Manufacturing Chemist, June 1994, pp17-19

8 CONCLUSION

This short book has been about risk assessment and risk management as applied to the pharmaceutical sector and healthcare. The book was designed to introduce the topic, provide details of the current legislation, and then to illustrate the concepts through some case studies.

One important point made in the book is that in operating risk management processes and undertaking risk assessment a core team is required. This should be made up of production staff, engineers, microbiologists, and quality assurance personnel.

To re-cap the key principles outlined in this book are:

Risk management

Risk management is the identification, assessment, and prioritization of risks followed by coordinated and economical application of resources to minimize, monitor, and control the probability and/or impact of unfortunate events or to maximize the realization of opportunities.

Methods, definitions and goals vary widely according to the risk management method.

For the most part, these methods consist of the following elements, performed, more or less, in the following order.

1. identify, characterize threats
2. assess the vulnerability of critical assets to specific threats
3. determine the risk (i.e. the expected likelihood and consequences of specific types of attacks on specific assets)
4. identify ways to reduce those risks
5. prioritize risk reduction measures based on a strategy

Risk assessment

Once risks have been identified, they must then be assessed as to their potential severity of impact (generally a negative impact, such as damage or loss) and to the probability of occurrence.

Risk assessment is the determination of quantitative or qualitative value of risk related to a concrete situation and a recognized threat (also called hazard). Quantitative risk assessment requires calculations of two components of risk (R):, the magnitude of the potential loss (L), and the probability (p) that the loss will occur.

To understand risk management and risk assessment, it is important that the principles set out in Quality Risk Management (ICH Q9) are reviewed and incorporated. This book has attempted to show how these core principles can be used through the various case studies.

Whilst the book has focused on aspects of pharmaceutical manufacturing Quality Risk Management can be applied not only in the manufacturing environment, but also in connection with pharmaceutical development and preparation of the quality part of marketing authorization dossiers. It is a flexible tool.

PHARMACEUTICAL MICROBIOLOGY

This book has been produced by Tim Sandle. Tim Sandle operates a number of Pharmaceutical Microbiology resources, including:

A LinkedIn Page
A Facebook site
And a discussion blog called Pharmaceutical Microbiology Resource,
at: http://www.pharmamicroresources.com/